A KOSMOS BOOK

IN A HUNDRED YEARS

In a Hundred Years

Alternative Energy Sources in the Post-carbon World

Presented in layman's terms

Hanns Günther
(Walter de Haas)

Translated by Thomas D. Hedden, Ph.D., C.T.

With 26 illustrations in the text
and a color picture on the cover

IngramSpark

Cover: Attempts to split the atom on Monte Generoso of the Lugano Prealps. Double strings of insulators intended to insulate ten million volts and exploit the enormous electrical voltages of lightning in thunderstorms for scientific purposes. Drawing: W. Goertzen

Contents

Original German title: *In hundert Jahren. Die künftige Energieversorgung der Welt*, published in 1931 by Franckh Publishing Company for Kosmos, Friends of Nature Association.
Original German printed in Germany / Holzinger Co. publisher's printing house, Stuttgart.
Original German copyright © 1931 by Franckh Publishing Company, Stuttgart
ISBN-13 of English translation:
978-0-9987887-9-1 (ebook); 978-0-9987887-8-4 (paperback), 978-0-9987887-7-7 (hardcover),
English translation copyright © 2023 by Thomas Hedden
Second printing November 2023

1. The World without Coal [1]

In 1913 there was a large international conference of
geologists. Among the many reports that were
presented there, one of the most important was
devoted to the Earth's coal reserves. Numbers are
usually thought of as dry. Here once again it was
shown that they can also be exciting, because these
dry statistics concerning coal production and
deposits suddenly raised the specter of a future coal
shortage. Initially, the numbers sounded quite
comforting. The report said that at our current rate of
coal, consumption known reserves of anthracite and
bituminous coal would last for another 6,000 years,
if mined down to a depth of 1,800 meters. The only
problem – continued the report – is that
unfortunately we are unable to utilize the entire
quantity, because many of the coal seams are so thin
that extracting the coal using today's mining
techniques is impossible or not worthwhile. Then, a
considerable portion of the coal that can be mined is
lost as dust during extraction. And finally, a large
quantity of coal is left behind in the mined seams,
since in many mines pillars of coal have to be left
standing to strengthen the ceiling. If we take these

---

[1] Translator's note: This book was written before natural gas
came to be widely used as a fuel and before the discovery of
the elephant oil fields in the Near East. If we imagine for a
moment that he is including these fossil fuels in his
discussion of coal, his points are clear.

circumstances into consideration and also the fact
that the Earth's coal consumption is increasing every
year, then it is out of the question that coal supplies
will last six millennia. Instead – concluded the report
– in the most favorable case the usable coal that is
known to us would most probably last us another
1,500 years.

However, this is not everywhere, but rather on the
average. This is an important, even decisive point.
The largest reserves lie in the United States; there
they will probably last another two millennia.
England presumably will already be out of coal in
200 years, and German deposits will last about twice
as long, at best. All these estimates depend on an
unknown factor: the constant increase in coal
consumption, whose future trend is unknown to
anyone. The reason why is that coal is used as a fuel
and raw material in countless places in the world's
economy. Coal is used so heavily and in so many
different ways that a world without it is hardly
conceivable.

Let us try to imagine for a moment what a "world
without coal" means. Above all, no more heating and
no way to cook. City gas could not be used either,
since after all it also comes from coal, nor could
wood, because our ovens would consume all existing
forests in a few years. And then, of course, railway
traffic and shipping would stop. But also automobile
traffic and aviation, because by then the Earth's
petroleum wells will long have been exhausted, and

all other synthetic fuels come from coal.
Furthermore, for most of humanity "No more coal!"
would mean: no artificial light, because there would
no longer be any city gas or lamp oil, and in most
countries electricity is produced from coal, so we
would be back to using fatwood and whale oil lamps.
However, all this would only be the beginning,
because naturally all industries that rely on coal as a
heat source or raw material would also come to a
standstill. For example, the large chemical industry
is based to a significant extent on coal and coke
products. And, most importantly the smelting of iron
ores is not feasible without coal, nor is the further
processing of pig iron into steel and malleable iron.
Without iron, the entire tool and machine-building
industry will fall. However, it would also fall just
due to the lack of heat and energy, since it also
derives these from coal. The lack of tools and
machines in turn means no modern agriculture,
among other things, and this would mean no food for
most of humanity.

Thus, "no more coal" amounts to the disappearance
of civilization! And, since no people would willingly
put up with such a downfall, the first consequence of
an incipient coal shortage would be that the coal-
poor countries would wage war against the coal-rich
ones. And also that both the coal-poor and coal-rich
countries would wage war against countries with
hydropower, which gives them an eternal annuity, in
contrast to the countries possessing coal, which are
consuming their capital.

However, there have always been opposing voices that have called the statistics into question, since we are always finding new coal deposits, and since there are enormous areas where a prospecting drill has never yet scratched the ground, especially in Africa and Asia. These objections are partly justified. Nevertheless, they do not change the fact that the Earth has only limited coal capital. And since this capital is constantly being consumed, sooner or later it must run out. Even the possibility of new coal formation has been dragged into the debate. However, this should not at all be expected, since it took this process millions of years to form what has been consumed in centuries. And the transitional stages from plants to coal that we possess in the form of deposits of peat and brown coal, as important as their occurrence might be for individual peoples and countries, contain so much less energy than the final product does that they will never be able to cover the needs of human development. – The coal problem also cannot be effectively addressed by conservation, that is, by using less. It is true that today enormous quantities of coal are still being wasted, especially by being burned directly in households. But even assuming that all coal would first be turned into coke, and that only coke would be burned everywhere, this would only stretch the reserves: it would still not make them any larger. This thought makes it obvious that there is only one really workable solution: today our coal capital is like a candle burning at both ends, since it is being used

both for energy production and as a raw material.
Energy production is consuming it most quickly by
far. If used only as a raw material, the Earth's coal
supplies would suffice for countless millennia. Thus,
it is necessary to eliminate coal from our energy
supply, and to replace it as quickly as possible by
other sources of energy.

If one thinks more carefully about the problem of
replacing coal, the first thing that automatically
comes to mind is hydropower, which, after all, long
ago displaced coal in countries that are poor in it, for
example Switzerland and Scandinavia. Every falling
or flowing body of water provides the ability to do
work, and the total energy of all hydropower on
earth has been estimated at over 450 million
horsepower. If we compare this figure with the
approximately 1,000 million metric tons of the
Earth's total coal production that is currently used
for power generation, which is equivalent to around
250 million horsepower, then any further
consideration appears pointless: "white coal" can
certainly replace black coal outright, if necessary!
However, unfortunately this simple calculation does
not stand up to closer examination. First, after all a
large part of the available hydropower is already
today being converted into machine power, i n
a d d i t i o n t o the power provided by coal. Of
course this portion cannot serve as a "replacement".
However, even more significant is the fact that
precisely the most powerful sources of hydropower
are located in areas that are hardly developed, far

away from all civilization and industry. Of the total quantity, around 190 million horsepower are in the remotest parts of Africa, where they flow away entirely unused. Another 54 million are hidden in the almost inaccessible jungles of South America. And another 70 million are in Asia. Harnessing these 314 million horsepower of hydraulic energy for humanity is, of course, not a technical but rather only an economic problem. But the further task of now delivering the electricity that is produced to the places where we need it, given today's distribution of industry and forces, is exclusively of a technical nature, since the means known to us today are totally insufficient to transport the electricity over thousands of kilometers. However, perhaps this task will already be solved in the near future. The reason why is that only one thing is needed: to learn how to control voltages of a million volts safely, not only in transformers and switching devices, but also in overhead power transmission lines.[2] And all prospects for this are present, since high-voltage lines for 350,000 volts are already in operation. But perhaps in a few decades we will even master wireless energy transmission, making all electric power grids unnecessary. What we see today as radio technology and broadcasting are the first timid steps on this road.

---

[2] Translator's note: It is interesting to note that the need to transport electric power over great distances still remains an important obstacle to increased use of alternative energy sources.

Therefore, we can safely assume that we will one day succeed in placing the world's entire 450 million horsepower of hydraulic energy at our service. Will this then finally satisfy the hunger for energy? Unfortunately, the answer is once again "No", since the 250 million horsepower of coal energy mentioned above that need to be replaced only corresponds to the power that humanity needs t o d a y . However, experience shows that the world's demand for energy doubles about every 20 years. Thus, by 1950 industrial companies will already be using 500 million horsepower, and by 1970 they will already be using a billion horsepower. This does not even take into account the fact that the 32 million motor vehicles on the Earth according to the latest statistics will need another 200 million horsepower, assuming an average power of only six horsepower. Today, these automobiles are being powered by petroleum wells. However, in the foreseeable future their power will have to be obtained elsewhere, since in all probability the petroleum wells will only be productive for another 2-3 decades.[3] Thus, wherever we look, we see that the world's enormously increasing demand for energy cannot permanently be covered from the available sources of power. The total available hydropower is far from sufficient to do this, even if it is completely used. Thus, the hope of finding help here in a future shortage is an illusion.

---

[3] Translator's note: This was written before the discovery of the "elephant" oil fields in the Near East.

2. Plans around the Mediterranean

Nevertheless, the solution might lie in hydropower, since there are many hundreds of millions of horsepower that have not yet been included in the statistics, which refer only to inland waters. For example, every second 88,000 cubic meters of water flow through the Strait of Gibraltar, from the Atlantic into the Mediterranean. If the level of the Mediterranean was lowered by 200 meters, this could provide a good 160 million horsepower: more than a third of the world supply. This possibility was first pointed out by H e r m a n n   S o e r g e l, whose Panropa project (also known as Atlantropa) is currently being intensely discussed everywhere. Soergel himself occasionally explained how he came upon his idea: not from the energy perspective, but rather out of economic policy considerations! The Panropa project is intended to raise and unite the economies of the crumbling West by requiring all the peoples of Europe to participate in a large, peaceful joint construction project. The goal of this work is supposed to be to exploit the dormant water power in the Mediterranean and to gain new land on the Mediterranean coast and in North Africa. In this way, Soergel wants to draw Africa close to Europe as a source of raw materials and a sales area, and create, from the two of them, a new part of the world – Panropa – to serve as a wedge between the Americas, which are financially strong, and the future Panasia, which is strong in population. We will briefly address the political side of these ideas later. But first, here we are most interested in the project's technical foundation, the way in which it is supposed to be realized. Soergel bases his idea on the fact that perhaps 50,000 years ago the Mediterranean's water level was around 1,000 meters lower than it is today. At that time, he says, approximately half the area that is covered by water today was dry land, fertile, inhabited, and probably the cradle of the early Central European civilization. Consequently, at that time Europe, Asia, and Africa formed a cohesive block. Then came the Ice Age, and after that the "Flood", a terrible diluvial disaster that turned millions of square kilometers of land into seafloor. It gave the Mediterranean the size and shape that it has kept since then.

What Soergel wants to do is basically nothing other than reverse this work of bygone millennia, and accomplish this using the powerful means that modern technology makes available. His plan relies on the fact that the Mediterranean loses a total of 4,144 cubic kilometers (km³) of water every year simply because the strong solar radiation causes evaporation that pumps a 1.65 meter deep layer of water off its two and a half million square kilometer surface year in, year out. Thus, the Mediterranean is an evaporation sea. Its level would automatically drop by 1.65 meters every year, if there was not constantly water flowing into it from outside that exactly balanced the loss. The largest part of the inflowing water comes from the Strait of Gibraltar, which is not, like many other waterways, a standing body of water but rather brings 2,762 km³ of water into the Mediterranean every year from the Atlantic Ocean. The same thing goes for the Dardanelles, except that the amount flowing in from the Black Sea is much less, only 152 km³. Another 230 km³ is supplied by the rivers emptying into the Mediterranean basin, especially the Rhône, the Po, and the Nile. The rest, which is around 1,000 km³, is provided by the annual precipitation over its surface. This list inevitably leads to Soergel's project: It involves closing the Strait of Gibraltar by a dam between Tarifa and Tangier (see Figure 1). This will stop the flow from the Atlantic. The Dardanelles will be closed in the same way by a dam at Çanakkale. The consequence will be that the level of the Mediterranean will be drop by about one and a half meters every year due to evaporation, which now has nothing counteracting it. To accelerate the lowering, Soergel also wants to use pumping stations and irrigation canals to connect the Mediterranean with low-lying areas of the Sahara (near Gabes; at the Gulf of Sidra; and in the Qattara depression), putting them partly under water. This would allow around six million square kilometers there, six times the area of France, to be transformed from useless desert into fertile land, as envisioned in the old Sahara Sea project. In addition to this, land would be gained in the Mediterranean basin, since Soergel wants to continue lowering the sea level until it lies 200 meters lower than it does today. This would cause the Adriatic to disappear almost completely, would cause Sicily to be joined with Italy and Sardinia with

Corsica, and around the coastlines areas many kilometers wide would emerge from the sea. This would produce a total of around 660,000 square kilometers of new land (see Figure 1), corresponding to the area of Austria, Hungary, and Germany including Danzig (Gdańsk) and the Saarland, and this would be land that could be settled, with rich and fertile soil, while most of the current coastlines are rocky and infertile.

Figure 1. Map of the Panropa project: The black areas indicate the new land (660,000 km²) that would be gained through lowering the Mediterranean by 200 m. The crosshatched areas indicate the desert areas in North Africa (6 million km²) that would be opened up to cultivation by irrigation

However, all that is only one side of this geological restoration. It goes hand in hand with an energy generation project that is even more magnificent from the technical perspective. On both sides of the dam in the Strait of Gibraltar, bypass canals are to be excavated (see Figure 2). These canals would carry the 88,000 cubic meters of water that today flow from the Atlantic into the Mediterranean every second and feed it to enormous hydroelectric power stations to exploit the difference in the water level across the dam to generate electric power, and simultaneously to return to the Mediterranean the quantities of water that evaporate, as soon as the desired lowering has occurred. Soergel calculates that a 200-meter difference in level would make it possible to produce 160 million horsepower, a thousand times the power produced by the power station on the Walchensee in Germany. However, of course the commissioning of the power stations would not wait until the lowering was completed, since after all this would take several decades. Rather, some of the power plants would be put into operation as soon as possible, perhaps after the Mediterranean is lowered by 30 meters, to produce the electric power to operate the Sahara pumping stations and the earth moving machines. As the level is lowered, the power generated would continue to increase. The peak power, including the 7.2 million horsepower from the Dardanelles, which would be set up the same way, would suffice not only to supply light and power to the new settlement land, but also to provide a considerable amount of power for other purposes.

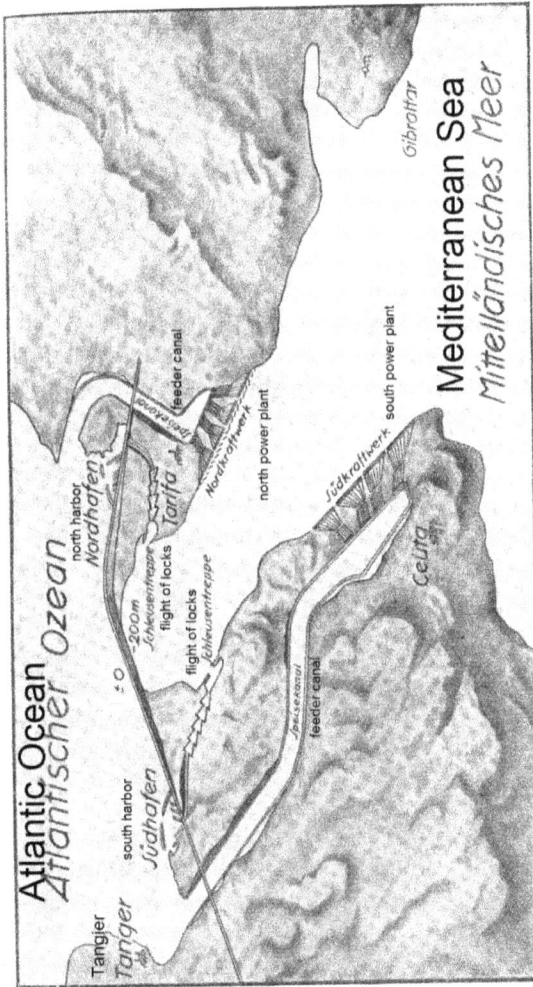

Figure 2. The Strait of Gibraltar, with the 29 km-long dam proposed by Soergel, which is intended to block the flow into the Mediterranean and simultaneously make it possible to produce 166 million horsepower in two giant power stations

So much for the idea, which could excite Jules Verne. But what about its feasibility? Soergel wants to build the large Gibraltar dam not at the narrowest place, where the sea is up to 500 meters deep, but rather in an arc between the Bay of Tangier and the Cabezos reefs, a stretch with many shallow places whose greatest depth is 320 meters. This would make the dam 29 kilometers long, so that if its base is 550 meters wide and its summit is 50 meters wide it would require around 10 billion cubic meters of earth, which Soergel wants to get from excavating the side canals. However, of course this is not done with a simple earth dam; that would only be sufficient with bodies of water at the same level. Here the dam must be watertight and able to withstand pressure, since after all it must separate the ocean from the Mediterranean, which will ultimately lie 200 meters lower. Thus, earth is only possible as a fill material. The outer layers must be made out of cast concrete or masonry, exactly the same as other dams are. Here this calls for underwater work, that is, immense caissons and diving shafts. It seems questionable whether engineering is yet capable of coping with this task today. But in a few decades we certainly will be. Of course the critical point is the joining place for the two halves of the dam, which will be built out from both sides. Closing the final opening will be immensely difficult: after all, it will be resisted by the entire force of the powerful current that now flows from the Atlantic to the Mediterranean. All the same, this is ultimately also only a technical problem that can be solved somehow, since such tasks have already been mastered on a smaller scale. This is not the case with other difficulties, for which reason they appear more important. The Mediterranean has numerous harbors, some of which are of world significance (Marseille, Genoa); lowering the level of the Mediterranean would make all of them unusable, since many kilometers of land would separate them from the coast. And how will the Earth behave if the water load borne by the present coasts should disappear? After all, the floor of the Mediterranean is not at all stable; this is already shown by Italy's volcanoes, which have been active for millennia. These questions are of much greater importance than the technical difficulties, since all the other details envisioned by the project, namely locks for ships on both sides of the dams; a 73 kilometer-long flight of locks at the entrance to the Suez Canal, power stations at the mouth of the Rhône and the Nile, machine units of an unprecedented size, power lines many thousand kilometers in length, removal of the salt covering on the drained areas, reclamation of the untouched soil – are tasks that engineering will be able to achieve in a few decades. And even the required capital, which is initially estimated at eight billion dollars,[4] probably would gradually be raised, since a good return seems assured. The only unknown factors that remain are nature and … humans. Soergel openly says that the ultimate goal of the Panropa project is to unite Europe and Africa into a powerful part of the world between Panamerica and Panasia. Fusing Europe and Africa and creating land bridges between them by partly draining the Mediterranean and irrigating parts of the

---

[4] Translator's note: This would be equivalent to over 100 billion dollars today.

Sahara means nothing less than eliminating the existing power structures on the Mediterranean and replacing them by a higher one that is more comprehensive, one in which all of Europe's civilized nations can participate. This goal, which has extraordinary significance for the world economy, involves numerous sweeping changes to existing political relationships. England's dominant position in Gibraltar would suffer; Italy would easily double in size. All this would depend on a dam which, if blown up during a war, would simply flood areas bigger than Germany. Who would protect the involved countries against this possibility? Who would protect the irrigation canals and the high-voltage lines that pass through so many countries, crossing dozens of borders? Merely asking the question already provides the answer to it: A prerequisite for carrying out the Panropa project is a complete change in Europe's entire spiritual makeup, complete elimination of the old Europe that is divided into many nations and just as many politically opposing forces. Until that happens, the ingenious plan, although it appears technically feasible, is no more than a utopia. But in a hundred years, perhaps it will long ago have become a reality, as the first great joint act of the United States of Europe.

*

The Frenchman Pierre Gandrillon based his project to generate energy from the Mediterranean on

completely different considerations. Its scope does
not even approach that of the Panropa plan, but its
technical foundation is better, since it is already
entirely feasible today. Then it has the advantage that
it can be repeated in numerous places on the Earth:
anywhere the natural conditions exist. The physical
basis is very simple and easy to understand. Let us
imagine two water containers: one at sea level, and
the other on a mountain. Then, by connecting the
two we can produce a hydraulic head that is as just
as inexhaustible as that of a natural waterfall, and
just as suitable for power generation, if two
preconditions can be satisfied: the higher container
must be inexhaustible, and the quantity of water fed
to the lower basin must disappear from there at the
same rate.

Whoever thinks over the possibility of such a
perpetual motion machine very carefully will soon
realize that these seemingly absurd conditions can be
realized in many places. In fact, they already have
been realized thousands of times, since our rivers are
based on precisely the same mechanism. For them,
the high-elevation, inexhaustible container is the
atmosphere, whose precipitation feeds all springs;
and the lower-lying basin from which the quantity of
water fed into it keeps disappearing again is the sea,
which, when heated by solar radiation, acts like an
enormous natural boiler, sending the entire quantity
of freshwater fed into it back into the atmosphere in
the form of vapor. Thus, in the end our perpetual
motion machine simply comes down to the idea that

since natural hydropower is insufficient, humans should themselves act as creators and remodel the Earth according to their wishes, to create artificial new waterfalls.

Gandrillon suggests reversing the natural process, which sounds fantastic, but his suggestion appears very rational if we follow it. Then, the role of the inexhaustible container is played by the sea. There is no doubt that it really is practically inexhaustible. In particular, Gandrillon first thinks of the Mediterranean, since its coastal areas in North Africa and the Arab world have numerous depressions that could be used as "lower-lying basins". As examples I would mention the Shatt al Gharsah in Tunisia (21 meters below sea level) and Melrhir in Algeria[5] (31 meters below sea level), then the area in Libya between Awjilah and Siwa (30 meters below sea level), but above all the Ghor depression in the Jordan Valley with the Sea of Galilee in the North (293 meters below sea level) and the Dead Sea in the South, which lies 394 meters below the level of the Mediterranean, so that here a hydraulic head of almost 400 meters is available. All these areas also meet the second condition, since they have very little rain and very strong solar radiation, and are characterized by very high temperatures[6] and

---

[5] Translator's note: The German original has "Schott Rharsa in Algier ... und Meb-Rir in Tunis"; the author apparently mixed up which one was in which country.

[6] Author's note: In the Jordan Valley the temperature is already over 40°C by the beginning of May.

correspondingly strong water evaporation. Thus, here the task of removing water after it is fed in is easily accomplished by the Sun, assuming that the water supply is correspondingly adjusted to match the evaporation capacity. Therefore, this is not really a perpetual motion machine. Even here, the energy obtained is counterbalanced by equivalent energy consumption, according to the law of conservation of energy. However, the energy that is consumed is not provided by us or by terrestrial means, but rather by the Sun, whose thermal radiation, which remains inaccessible to us for the time being, is made usable in a simple way by this plan. Thus, Gandrillon's plan is ultimately a solar power plant. But it is not one of the pathetic attempts that involve heating a tiny boiler by concentrated solar heat using expensive optics. Instead, Gandrillon uses the same process that drives the Earth's water cycle: evaporation of surface water from lakes and seas, which act as giant natural boilers.

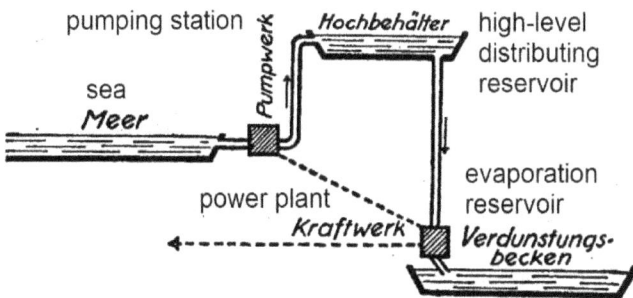

Figure 3. Project to produce energy from the Mediterranean in combination with natural depressions, schematic

However, here as everywhere Nature requires work before pay, since the project is not entirely free of difficulties. All depressions that are possible for the plan are separated from the Mediterranean not only by wide stretches of coastline, but also by elevations. Thus, the first thing that needs to be done is devise a practical and economical way to connect the two basins, despite this obstacle. To accomplish this, Gandrillon proposes the process illustrated in Figure 3. Brief examination shows that his basic idea doubtless comes from the large pumped-storage plants that our energy industry has recently begun using more and more in connection with hydroelectric and thermal power plants to store their excess electric power. Between the sea and the evaporation basin a third member is inserted into the system in the form of a high storage basin at the crest of the separating elevation. Giant pumps draw the water out of the sea and force it through ascending pipes into the intermediate basin. From here, it falls down through descending pipes onto the turbines of a power plant, and then goes into the large evaporation basin. If the drop is greater than the height difference that the pumps have to overcome, then only part of the energy produced is needed to drive the pumps, while the rest is freely available.

Gandrillon applies this idea to the conditions in the Jordan Valley, which at the latitude of the Sea of Galilee is separated from the Mediterranean by a strip of land that is around 50 kilometers wide. Its lowest point is 80 meters high and is approximately 20 kilometers from the coast (see Figure 4). This would be the place to set up the high storage basin to which the sea water is fed by powerful centrifugal pumps. From here – that is from a height of 80 meters above sea level – the water is supposed to go through a long, partly underground canal to a surge tank on the other side of the mountain, and then fall down almost 300 meters through descending pipes to power station I near the Sea of Galilee, 208 meters below sea level. Once the water has done its work in the turbines, it is carried through an open canal on the other side of the Jordan river bed into the vicinity of the Dead Sea, where the remaining drop – another 180 meters – is exploited in a second power station. Its tail race is supposed to flow into the Jordan at the Jordan River ford.

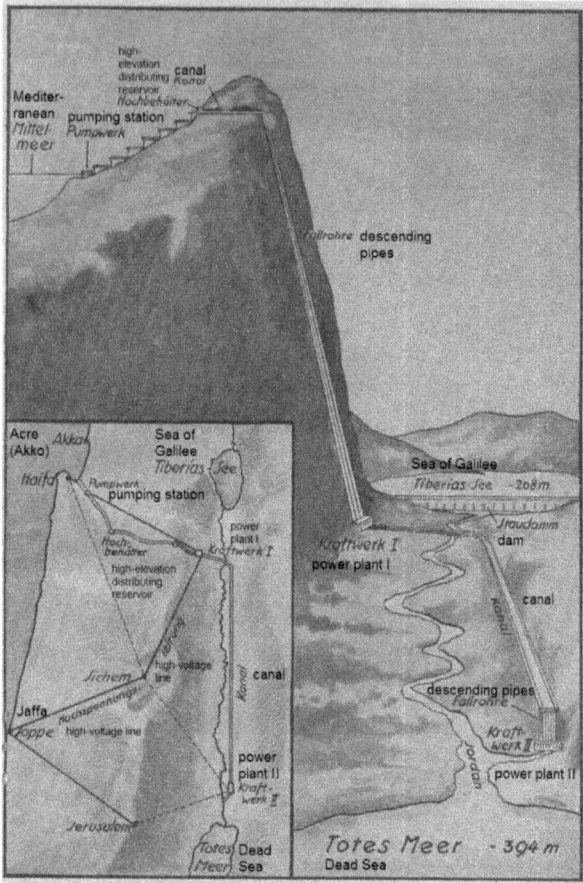

Figure 4. At the top left is the Mediterranean, whose water is raised by a large pumping station into a high storage basin. From here, it falls down through canals and descending pipes into the Jordan Valley, an enormous artificial waterfall whose drop is exploited in two power stations.

To see what power can be produced in this way, we must first know the evaporation capacity of solar heat. In French salt works, the Sun evaporates a 16-millimeter layer of water every day. However, these basins are very shallow; for large water depths no more than 3-6 millimeters can be expected. This corresponds to the evaporation balance of the Dead Sea, since around 100 cubic meters of water currently evaporate every second from its 926 square kilometer surface. However, this evaporation is not visible, since it is exactly replaced by the inflows – above all the water from the Jordan. If we start with these conditions, then it follows that adding water from the Mediterranean must cause the level of the Dead Sea to rise, e.g., given an inflow of 30 cubic meters per second, we calculate that it would rise one meter in a year. However, this calculation assumes a container with vertical walls, while on the North and South the Dead Sea is bordered by enormous desert areas. Of course the rising water would spread out over these areas. This would considerably increase the evaporation surface and would cause the evaporation to increase by a corresponding amount, namely by 11 cubic meters a second for every 100 square kilometers of new water surface. If the water inflow is adjusted accordingly, then using relatively modest means it is possible to produce 20,000 horsepower, without making the slightest change in the current water conditions. However, an amount of power this small would be of only local significance. Therefore, Gandrillon's plan is much more ambitious. He wants to dam up the water of the Jordan in the Sea of Galilee, and also divert the other tributaries of the Dead Sea to irrigate areas that cannot currently be cultivated, due to their great dryness. This would make it possible to increase the inflow from the Mediterranean enough to allow the power stations to supply around 250,000 horsepower. Finally, it would also be easy to keep the water inflow higher than the evaporation figure for a few decades. Then the Dead Sea would rise by a certain amount every year, which would not harm anyone. But the power that could be harnessed would double.

Finally, it should also be noted that the areas around the edge of the Mediterranean are by no means the only possible places where this concept can be realized. There are many other places on the Earth, for example the Caspian Sea (– 26 meters), Lake Assal on the coast of Somalia[7] (– 174 meters), Death Valley (– 84 meters), and the Coachella Valley with the Salton Sea (– 90 meters) in the United States have all the preconditions for such energy production. It follows from this that Gandrillon's project represents a very noteworthy contribution to the Earth's future energy economy. This is especially true if one follows the example of the Panropa plan and combines this project with the obvious idea of

---

[7] Translator's note: The German original has "der See von Assal an der Somaliküste"; it is uncertain whether he means Lake Assal in what is now Djibouti or Lake Karum in the Afar Region of Ethiopia, which is also known as Lake Assal.

creating enormous inland seas in hot areas that cannot be settled today, in order to change the climate there completely and turn deserts into fertile land. The energy economy of the future will certainly consider such combinations important, if they are possible, just for their profitability.

3. Will There Be Wave Power Plants?

Engineering always offers many different approaches to every question. This is also true in the case of alternative energy. Our beautiful planet offers a number of other energy sources aside from coal and water power. Thus, the best thing is to determine which provides the most energy. The obvious inventory can quickly be surveyed. If we disregard coal, petroleum, rivers, and waterfalls, then the following are possible:

the infinite quantities of solar heat that cover the Earth;
the enormous glowing ember inside our planet, a dowry from the Sun;
the atmospheric airflows that we call winds;
the waves that break on the coasts of the seas in a thundering surf;
the never-ending rise and fall of tides, the periodic change in sea level commonly called ebb and flood tides caused by the Moon's gravitational attraction.

Solar radiation, winds, and geothermal power will be
discussed later. Here let us first say a few words
about o c e a n   w a v e s . Their energy content
appears above all where they pound against rocky
coasts as a roaring surf, strong enough to seize and
smash large ships. At times, such surf waves contain
absolutely inconceivable force. The Englishman
Stevenson attempted to measure them, and found
pressures of 15-35 tons per square meter of coast.
Carrying this calculation further shows that when
there is a single powerful gust of wind, the west
coast of France, which is known for its strong surf, is
hit by a wave whose kinetic energy is equivalent to
100 million horsepower.

But there is no point in marveling at such numbers,
since they are only instantaneous values that cannot
always be counted on. Their source is too capricious,
too erratic. The place where waves unleashed by the
whip of the storm break through levees in a sudden
rage and wash away entire villages today can have
calm water with ripples playing over the surface
tomorrow. And the same rocky beach that on one
day shakes under the roaring of the raging waves
might invite you to go swimming the very next day.
This variability is the most difficult obstacle to
exploiting the ceaseless play of the waves, since the
most important conditions for industrial utilization
are long duration and steadiness.

Nevertheless, there continue to be experiments with
wave motors. However, most of them are of a very

primitive nature, since they confine themselves to the most obvious idea of using floats that go up and down. This movement is transferred by levers to a transmission, and from there on to pumps, which work to fill an elevated storage basin with sea water. The water that is raised is used to drive turbines, producing electrical energy. Instead of floats, other approaches use vertical boards that the waves strike and cause to oscillate like pendulums. This oscillating motion is then transferred by a simple lever system to compressed air pumps. This system is intended to fill large tanks with compressed air, and these "reservoirs" drive generators.

All this sounds entirely plausible on paper. Until one begins to do calculations, that is. Every transformation of one motion into another consumes power. And here there are several successive energy transformations. The consequence is very low efficiency. This, combined with the unsteadiness of the initial source of energy, makes it obvious that such proposals are unusable. That may be true, but their authors say that wave energy is unlimited and inexhaustible. Thus, they feel that even such low efficiency does not play a role. Moreover, they say that one should imagine such arrangements on a very large scale. For example, the floats would be heavy ship hulls, dozens of which would be anchored next to one another at places where waves are strong, and their up and down motion would be transferred to a transmission through giant bridge-like lever apparatus. Such "machinery" actually would yield a good number of kilogram meters. The only problem is that making the system so large immediately shows another weakness of the process: building the connection over to the land so that it is not too heavy, but can still bear its load. Given the enormous forces that follow from Stevenson's measurements (see page 30), this is an inauspicious beginning.

However, there is another possible way to extract the energy of ocean waves. It forms the basis of the American project illustrated in Figure 5, obviously the only one that was designed by someone with some expertise. In order to understand it, we must digress a little. It is best for us to start from a well-known phenomenon: the sharp bang that is often heard when a water faucet is very quickly closed. This bang is due to the fact that the flow of water that is suddenly stopped by closing the faucet can only release its energy as a shock against the walls of the pipe. Such a shock can be strong enough to burst the pipe. However, if the pipe has a place near the faucet that is not watertight, then the recoil manifests itself differently: the moment the faucet closes, the water squirts out of the pipe in this place in a tall stream. Brief reflection makes it clear that this phenomenon can be used to raise water. This has already been known for more 100 years, since in 1796 Joseph Montgolfier, one of the two brothers who sent up the first hot-air balloon, used it as the basis for constructing an automatic water raising machine that is called a water hammer or hydraulic ram. Figure 6 schematically illustrates the device. A rather large-diameter descending pipe supplied from a large water reservoir is closed at the lower end by a valve. The valve body V is heavy enough that the pressure of the s t a t i o n a r y water column is insufficient to raise it. Thus, the valve remains open, and the water flows out. However, after a short time the flow velocity, which is rapidly increasing, becomes great enough that the water pushes the valve body up. This closes the valve and stops the outflow. The consequence is a strong recoil in the descending pipe, opening valve Z in the wall of the descending pipe. Then, part of the water flows through this opening with great speed, i.e., great momentum, into tank W. This compresses the air here, and drives the water into the ascending pipe leading vertically upward. Once the recoil has released its energy in this way, valve Z closes again under the weight of the water resting on it. Simultaneously, valve V opens, and thus the cycle starts all over again. It repeats as long as water is flowing out fast enough at V. Air tank W evens out the pressure differences exerted on the vertical ascending pipe as a result of the pulsating operation. Consequently, there is a uniform flow of water out of the top end of the ascending pipe.

Figure 5. American design of a wave power station
based on the principle of the hydraulic ram

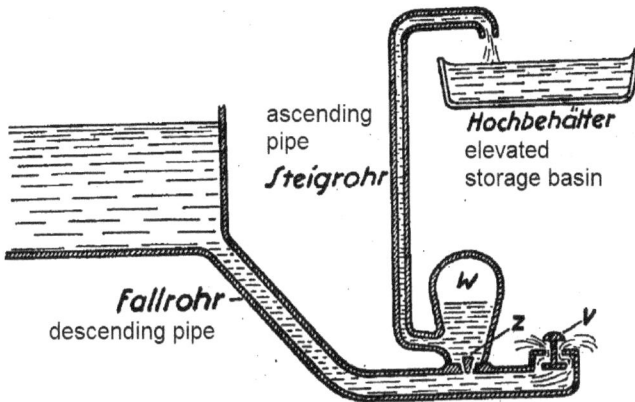

Figure 6. Schematic illustration of the mechanism of action of the hydraulic ram

This information makes it easy to understand Figure 5. Here the container supplying the descending pipe is replaced by the sea, and the water pressure to produce the necessary flow velocity is supplied by breaking waves. The large floating funnels are wave catchers that convert the surf into a current. The recoil of this current pushes the water into the ascending pipes, which convey it into a high-elevation reservoir. Its fall feeds a hydroelectric power station of conventional design. T h i s r e a l l y i s b e s t w a y i t c o u l d b e d o n e ! The unsteadiness of the breaking waves is evened out by putting the reservoir in between. And if a big enough margin is created between how much water is raised to how much is consumed, continuous operation of the power station is ensured. Many mechanical difficulties are also eliminated here, since after all there are not any complicated transmission parts. It is another question how expensive such systems would be. If designed in sufficiently large dimensions, they would probably be so expensive that economical operation would be impossible. The only circumstance under which economical operation would not play a role would be if, in the distant future, we absolutely needed ocean waves as an energy source. However, the discussion below will show us that there is no way that this could be true. There are other possibilities that are easier to develop and that give better results. Therefore, ocean waves do not deserve a place in a serious consideration of our energy problem.

4. Ebb and Flood Tides

Every 24 hours, the Earth rotates once about its axis. At the same time, it is circled once by the moon. The moon describes this orbit because the Earth's gravity acts on it. And, conversely, the Earth is acted on by the gravity of its satellite. This is apparent mainly in places where the Earth's surface is covered by the oceans, since the easily moveable water yields to the Moon's gravity much more readily than solid land does. The consequence is the phenomenon known as the tides, illustrated by Figure 7 [8]: on the Earth's side

[8]) Author's note: For simplicity, Figure 7 assumes that oceans cover the entire surface of the earth. The actual conditions cannot be represented on such a greatly reduced scale.

A directly facing the Moon, the ocean rises. However, it also rises on the side D that is facing away from the Moon. Here the Moon's gravity acts more strongly on the solid land lying closer to the Moon. Thus, the ocean has the ground pulled out from under it, so to speak. The result is the same as in case A: The water "swells", it rises with respect to the land. This rise of the ocean with respect to the land is called a flood tide. Its necessary counterpart is the ebb tide: at points B and C, which are 90° apart from points A and D, the ocean falls when it is rising at A and D. However, since the Earth rotates once about its axis every 24 hours, after six hours point A will lie where B lies now. in other words: six hours later A has an ebb tide. However, in the meantime B has also moved by 90°, so that it occupies point D, and has a flood tide. After another six hours, A has rotated by 180°, so it has arrived at D, and once again has a flood tide, while B lies at C and once again has an ebb tide. Points C and D move in the same way, as do all other points on the Earth. The result is that every ocean-covered place on the Earth has two ebb tides and two flood tides every 24 hours. The ebb tide manifests itself by the water flowing away from the coast and leaving the ocean floor dry, while the flood tide once again covers the strips of land that were previously left dry. The movement of the Moon about the Earth and the division of the Earth's surface into continents and oceans do not, on the whole, change the nature of the tides. However, they do cause changes in detail and at individual places. For example, the tides have almost no effect in the Baltic Sea or the Mediterranean, which are connected with the open ocean only by narrow waterways. In other places, the difference between flood tide and ebb tide is normally about three meters. However, there are also coastal areas whose location and terrain features are especially favorable to the formation of the advancing front of the flood tide, so that it reaches heights of 10 and 12 meters. In Europe, the west coast of France and the southwest coasts of England and Ireland have the best prerequisites. They are found above all in fjord-like ocean bays and long, deep-cut river estuaries. In such places, a spring tide [9] can produce "water mountains" of up to 16 meters high that flow inland with increasing speed, and afterwards flow back towards the ocean at even greater speed. When there is a spring tide, the rising lasts five hours and the falling lasts seven and a half hours; when there is a usual flood tide, both processes can be expected to take six and a half hours.

---

[9] Author's note: We have a spring tide when there is a new moon or a full moon, since then the Sun, which also produces a weak tidal impulse, reinforces the Moon. – The counterpart of the spring tide is the very low neap tide, which occurs when the Moon is in its first or third quarter; in this case, the Sun is counteracting the Moon.

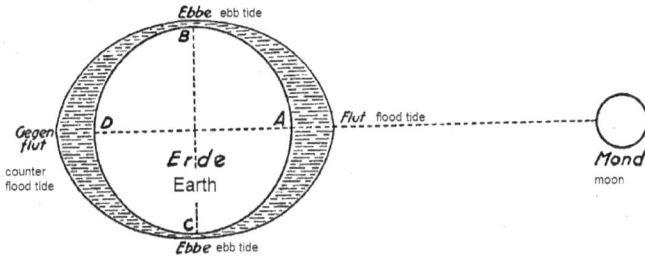

Figure 7. Schematic illustration of how tides form

Of course it is obvious that the movement of such large masses of water involves an enormous amount of energy. Attempts to calculate its value, at least approximately, show that the tides are the second-largest energy source available to us, after solar radiation. Its annual mechanical equivalent is estimated at the inconceivable value of eleven quintillion horsepower. In comparison, the billions of horsepower years that humanity will probably need to satisfy its demand for energy in 50 years is vanishingly small. But this tempting treasure is like the gold in seawater. It is there: 5,000 kilograms in every cubic kilometer! Only we cannot get it, since the gold is dissolved in it at an infinitesimal concentration. The situation is quite similar in the case of tidal energy. It is present, but only the smallest part of it can be exploited! The reason why is that doing so requires quite specific conditions that are only found at a few places on the Earth. The main difficulty is that the amount of available energy fluctuates very strongly. The flood tide begins to rise slowly, gradually becoming stronger and stronger, and reaches its high point rather quickly around six hours later. Shortly after that the water begins to drop, slowly at first, later more and more quickly, until it completely ebbs after about six hours, and then the flood tide begins all over again. Superimposed on this twelve-hour up-and-down cycle are the deviations caused by the relationship of the Moon to the Sun (see previous page). Finally, there is also the influence of the wind, which can, if it drives the flood tide, lead to unusually high tidal waves – a storm surge.

The significance is obvious: Although the tides keep repeating, they are definitely not a steady process. And absolute steadiness is the most important condition for industrial exploitation of a natural force. In order to make the tides steady, it is necessary to dam up the flood tide in large basins behind dams. In such facilities, the available work depends simply on the surface area of the basins and the mean drop. Since the drop is always small, somewhat significant power production presupposes basins having a surface area of many square kilometers. Under these circumstances, flat sea coasts can be immediately ruled out for tidal power plants, since here the storage basins generally can only be created by dams that are kilometers in length, and their construction would be enormously expensive. But not even all of the steep coasts, few as they are, are suitable. Rather, the only practical places are those where Nature herself has created the storage basins in broad bays whose entrances are as narrow as possible, since then closing them requires only a relatively short dam. Deeply cut estuaries can also offer similar conditions. However, more often than not an obstacle is presented by the necessity of keeping a path open for navigation, since large locks enormously increase the cost of the work. Another important thing is for the storage basins to be able to fill and empty very quickly, to allow the inside and outside level to be balanced precisely in the rhythm of the tides. However, there is an obstacle to meeting this demand, namely the fact that the dam's strength only allows a limited number of passages to be built in it. And naturally these passages are needed to operate the turbines. But this number is not enough in practice, since the turbines can draw only limited amounts of water. Thus, it is necessary to create additional equalization installations for the excess water. Finally, it must still be possible to maintain the operation during the time when the turbines cannot work since the difference in elevation of the two water levels has fallen below a certain minimum value. On the average, these "slack water times" take up 2-3 hours of the 12½-hour tide period. There are various possible ways to continue to supply power during the pauses: a second storage basin as a reserve, or coupling with a pumped storage or steam accumulator power station. But such "auxiliary plants" are expensive, since after all they must be dimensioned so that they can completely replace the main plant. Thus, basically it is always necessary to build two plants. This necessity puts a great strain on the construction budget.

We see that it is not easy to use tidal power. Nevertheless, there is hardly another technical task like this one that has always been so alluring to professionals and amateurs alike! The first suggestions for tidal power plants that we know about go back to the 11th century. Since then, the literature about this has increased without stop. However, what is more important is that people keep trying to make the idea a reality with "flood tide mills" driven by simple water wheels, though at first only on a very small scale. Here again, it has been left for us to enlarge the scale. Currently England, France, Argentina, and of course the United States are seriously occupied with such projects, some of which are of a truly enormous size. Before World War I, Germany was also in this list, since around 1910 a Hamburg engineer, E. F. Peine, wanted to set up an "electric tidal plant" at Husum on the North Sea. This project was the first attempt to solve the problem of exploiting the tides using modern means. The natural conditions were favorable. Three sides of the "storage basin" were present: the island Nordstrand, the mainland opposite it, and the 2,800 meter-long railroad embankment connecting the two of them. However, taken as a whole, the project was modest: the power plant was supposed to have a capacity of only 7,500 horsepower. However, it would have been a beginning, and the experience gained would certainly have been led farther. Unfortunately, three years were spent in useless discussions. Finally, in 1913 a small trial power plant was built to settle certain disputes in practice.

The result was good, since all of Peine's predictions proved correct. The project already appeared to be a sure thing, and then World War I broke out. The experiments were halted, and the trial power plant was demolished. Since then, nothing more has been heard of these plans.

This is understandable, since the flat North Sea coast, with the relatively small height of its flood tides, is out of the question for a large power plant. And of course in practical terms, this is precisely where everything calls for power plants of ever larger capacity. There are excellent preconditions for this on the French and English coasts. Therefore, in both countries the government has also been devoting special attention to the tidal energy problem for a long time.

A few years ago, France built, at the mouth of the Diouris at l'Aberwrach (see Figure 8), an experimental power plant having a capacity of 2,500 horsepower, which is in operation an average of 10 hours out of the 12½-hour tidal period. During the slack water period, an auxiliary power plant located six kilometers upstream supplies power. It is a pumped storage power plant whose turbines exploit the drop of the dammed-up Diouris; however, it also contains three large pumps that are powered by the excess current of the tidal power plant. Thus, while the auxiliary power plant is idle, the water previously used by its turbines, which has been collected in a reserve basin, is pumped back into the storage basin, in order to increase its energy reserve. Both power stations are connected to a 30,000-Volt line, which transfers the generated current to Brest. There it is delivered to the Brest Arsenal and other government operations.

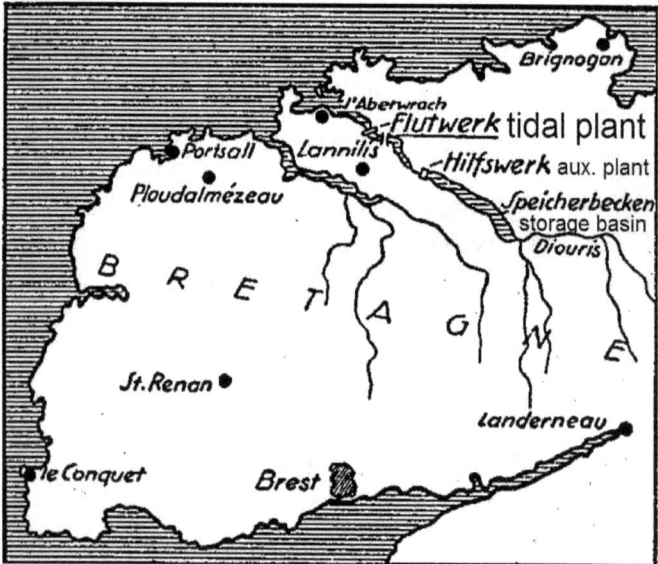

Figure 8. Site plan of the French experimental tidal plant at the mouth of the Diouris in Brittany

Whether anyone will want to evaluate this experiment on a larger scale someday is unknown. Although there are plenty of corresponding projects right in France, none of them seems to have been seriously studied up to now. This situation is different in England. There, a few years ago the Minister of Transport announced a plan to create a tidal power plant having a mean capacity of 500,000 horsepower at the mouth of the River Severn in the Bristol Channel. As is shown in the diagrammatic map in Figure 9, the Bristol Channel forms a fjord-like estuary that is about three kilometers wide where the dam is supposed to be set up. The tides are unusually strong at this place. Even when there is a neap tide, the advancing front of the flood tide still reaches a height of nine meters. The banks are very steep towards the sea, but flat inland. This is favorable for dam construction. The dam is supposed to be four kilometers long and close off a 70-square kilometer storage basin. The upper parts of this basin would simultaneously form a large harbor, for which the adjacent industrial district would have good use. Some of the construction costs could be borne by this harbor and the large lock forming its entrance. Another part of the construction costs could be borne by a road and railway bridge over the River Severn that has already been planned for a long time; this project presents an opportunity to build it. These details have been portrayed in the picture of the entire project shown in Figure 9. In the foreground, the sea is closed off by a massive dam, whose ground plan follows the natural shape of the terrain. The innumerable passages through it (sluices) can be opened and closed very quickly by electrically operated sluice gates [10]. Behind the dam lies the

[10] Author's note: In engineering, the term "sluice gate" designates the iron plates running in lateral guideways which can be used to block the openings of a dam, entirely or partly, as required. In large dams, the sluice gates are raised and lowered by special electric motors.

large railway bridge, which in the middle has a loop leading to two lift bridges. This arrangement is necessary to keep the operation of the large lock from interfering with railroad traffic. The machines to produce electricity lie inside the dam (see Figure 10), each sluice having a hydroelectric generating set consisting of a turbine and a generator with a mean power of 1,800 horsepower. There are 280 such groups planned. This gives total mean power of around 500,000 horsepower. Here the slack water also lasts an average of 2½ hours per tidal period. In order to continue to supply power during the pauses in operation resulting from the slack water, the project originally called for, as shown in Figure 9, a pumped-storage power station at Tintern, with a reservoir in the Wye Valley and machine units of 18,000 horsepower each, consisting of a turbine, a centrifugal pump, and a generator. At night, the generators were intended to run as electric motors powered by the excess current of the tidal power station, and drive the centrifugal pumps to pump the water of the Rive Wye into the reservoir. This water was supposed to be used to operate the auxiliary power station, which was also planned for a power of 500,000 horsepower, during the slack water periods to take the place of the main power station during its pauses in operation. However, this project seems to have been abandoned in the meantime.[11] The reason why was that the entire project was the

---

[11] Translator's note: Although the Severn Barrage project has been set aside from time to time, it was still being discussed at the time of this translation.

object of sharp criticism in England, mainly with regard to the estimated cost of 30 million pounds, which was declared to be too low, but also with regard to technical details. Among other things, critics pointed out that before undertaking such a colossal work, it is absolutely necessary to carry out basic research first to provide a scientific foundation for the problem of exploiting the tides in general, and in particular for the Severn. It was obvious that this demand was justified. And therefore the government agreed with it. To carry out the pilot studies, a small, experimental power station was set up in 1930 at Avonmouth Dock on the River Severn, however not with a pumped-storage plant as a backup for the slack water periods, but rather with an auxiliary steam power station that had an electrically heated boiler to exploit the excess current. The steam produced is fed into a large, thermally insulated container, and it is kept in this "steam accumulator" in order to feed a steam turbine during the slack water periods. It is used to drive the generator when the flood tide turbine cannot operate due to the lack of a drop. It can be seen that this is a quite simple, but precisely for this reason highly remarkable idea. If it proves itself, then this will represent considerable practical progress in solving the problem of exploiting the tides.

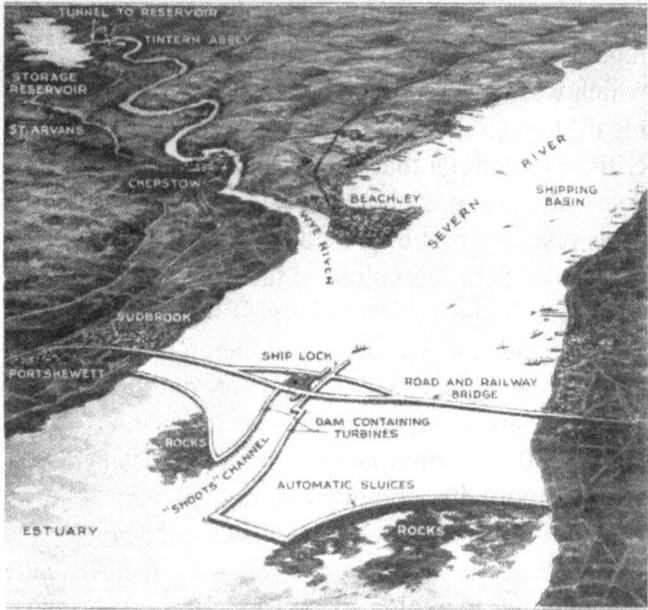

Figure 9. Bird's-eye view of the mouth of the Severn with the planned tidal power plant; site plan shown at top right [12]

Among the other tidal power projects, let us mention just one more, first because of its size, and then because it was the object of lively discussions at the second World Power Conference (*Welt-Kraft-Konferenz*, Berlin 1930). It was elaborated by an

---

[12] Translator's note: The figure shown here is the original illustration that appeared in *Popular Mechanics* rather than the one that was translated into German for publication in the German original of this book: Britain's Huge Tidal-Power Project. *Popular Mechanics Magazine*, vol. 35, no. 3, pp. 334-346. March, 1921.

engineering committee of the Argentinian government, primarily for the purpose of reducing the fuel-poor country's expensive coal imports. The prerequisites for tidal power plants are favorable on the coast of Argentina: The difference between high and low water is up to 12 meters, and south of Bahía Blanca all the way to the Strait of Magellan there are several estuaries and bays that can easily be closed by dams. Current thinking is that the most suitable is *Golfo San José*, whose position is shown on the schematic map in Figure 10. This bay has a surface area of 780 square kilometers, while its greatest depth is only 54 meters. Since the entrance is narrow, a dam seven kilometers long is sufficient to close it off, despite its enormous surface area. A lock for ships is unnecessary, since the adjacent *Golfo Nuevo* serves the same hinterland. The dam is supposed to be cast from concrete and extend another two meters above the highest water level. It is supposed to have several stories inside it containing the generators, turbines, sluices, and necessary workshops. The walls of the individual rooms facing the sea are supposed to be fourteen meters thick, and those on the opposite side are supposed to be four meters thick.

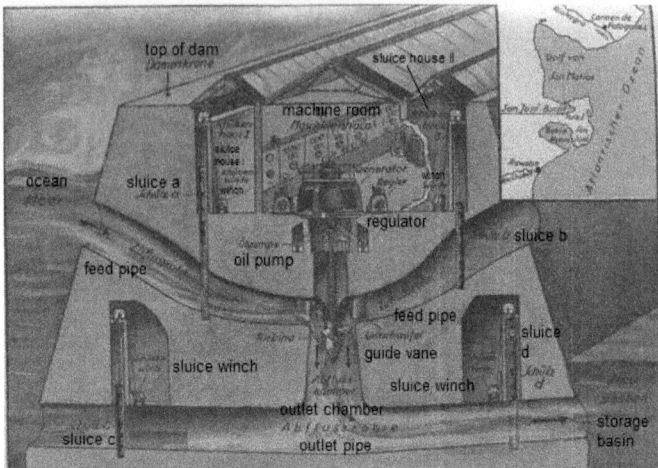

Figure 10. Cross section of the dam of a tidal power plant with the auxiliary systems and the power house built into the dam. Top right: Diagrammatic map of the tidal power plant planned by Argentina at *Golfo San José*

Figure 10 shows how we can imagine the rest of the system. The turbines are individually arranged in chambers with intake pipes and outlet pipes on both sides. All four pipes can be closed by electrically operated sluices. Horizontal turbines are used, whose shafts are vertical. The generator sits on the top end of the shaft. The water falls vertically down from above through the intake pipe into the turbine. Thus, the impeller coupled with the generator always rotates in the same direction, irrespective of whether the water is coming from the open sea when there is a flood tide or coming from the bay when there is an ebb tide. In particular, the turbines are built in such a way that they begin to run only once there is a 50-

centimeter difference in level between outside and inside, while they give their highest power when there is a drop of one meter. Accordingly, the water inflow is limited by automatic devices. In other words, if more water is available, the turbines only take in enough so that the most favorable drop between the sea and the bay is constantly maintained. However, this flow limiting has the consequence that the total intake capacity of the 376 turbines inside the seven kilometer-long dam is insufficient to even out the water levels of the bay and the sea completely during every ebb and flood tide. To even out the water levels despite this, shortly before the end of every ebb and flood tide the excess water is fed directly through the turbine outlet pipes. This produces suction, increasing the power of the turbines.

This information makes it easy to understand the essential features of the operation of a tidal power plant. Let us assume that the ebb tide is just ending. All sluice gates are open, but the turbines are not operating, since the water is at the same height on both sides of the dam. Now the flood tide begins. Upon a signal, all sluice gates are closed. The consequence is that the water on the sea side of the dam rises, while the storage basin maintains the ebb tide level for the time being. As soon as the drop is 50 centimeters, and thus sufficient to drive the turbines, sluice gates a and d are opened. This connects the turbine chambers with the sea, and the discharge chambers with the storage basin. Thus, the flood tide water now flows through the dam into the basin, and all machines begin to run. In the meantime, the sea rises further, and at first the water on the other side of the dam also rises on the same scale. That is, until the turbine reaches its maximum output. From that point on, the inflow regulators let only just enough water pass so that the drop between the inside and outside levels is constantly maintained at its most favorable level of one meter, corresponding to this output. With this drop, the machines continue to run until the flood tide almost reaches its highest level. Now, sluice gates c are opened to let in the last torrent of the flood tide. The storage basin is then directly connected with the sea through the discharge chambers; consequently, the outside and inside levels equalize quite quickly. The turbines continue to run as long as the drop is still more than 50 centimeters. The output is even greater than at the beginning, since during this period additional power appears in the form of the suction of the direct water flow (see above). However, this additional power is enjoyed only briefly, since the drop soon falls below the limit. Then, the turbines are shut down by blocking the water inflow and closing sluice gates a. Shortly thereafter, the water inside and outside is at the same height. This is the beginning of the first pause in operation, which lasts an average of 1¼ hours. Now sluice gates c and d are also closed, and the pause in operation continues until the ebb tide causes the water in the sea to drop by 50 centimeters. Then, operation can be resumed. To do this, sluice gates b and c are opened. The water then flows out of the storage basin through the turbine chambers into the sea. In this way, the turbines work about another five hours. Shortly before the end of this period, the basin is directly connected with the sea by opening sluice gates d, completely emptying it. This is followed by the second pause in operation of another 1¼ hours. Then, the flood tide begins again, and the whole cycle repeats.

It is obvious that enormous quantities of water flow back and forth in such a power station. Nevertheless, when one begins to calculate the numbers, their magnitude is surprising. At the San José plant, the water level of the bay would rise and fall an average of 2.6 meters during every tide period. Given a surface area of 780 square kilometers, this gives a total volume of two billion cubic meters of water that flow through the dam's sluices during every ebb and flood sequence. Coping with these volumes of water requires intake and outlet pipes 8-10 meters in diameter. The turbine output corresponds to these dimensions. The 376 hydroelectric generating sets that are distributed at intervals of about 19 meters over the 7,000 meter length of the dam supply a total of 1,000,000 horsepower given a drop of one meter, while the total output is still 400,000 horsepower given a drop of 50 centimeters. This results in a minimum daily current production of 15 million kilowatt hours, which can increase to double that during spring tide. If the annual production is calculated on the basis of the minimum number, this comes to around 5.5 billion kilowatt hours. The meaning of this is easier to understand if one hears that in the year 1929 all the Swiss hydroelectric power stations combined sold 5.52 billion kilowatt hours. Yet Switzerland is one of the countries with the heaviest relative current consumption. These quantities of current, generated by the work of the tides, would reduce the Argentine Republic's annual coal imports by several million tons. Therefore, the government is vigorously pursuing the project. At present, this involves building, in a side bay at the mouth of the Deseado River offering similar flood tide conditions, a small experimental power plant having a daily generation capacity of 1,000 kilowatt hours, which is intended to clarify all factors that are still unknown. Afterwards, it is desired to begin work on the large plan immediately.

Thus, there is no doubt that progress is being made everywhere on the problem of using the tides. Probably the next decade will already see the first large-scale tidal power plant come into being. However, in 100 years all suitable ocean bays will long ago have been developed. In many countries, the seacoasts will then form the supply centers for electric power.

5. Harnessing Cyclones.

We just described the wealth of energy available in the flow of tides as unimaginably enormous. However, even that is greatly exceeded by the unused energy in the Sun's thermal radiation, although the Earth actually receives only $1/250{,}000{,}000$ of the total radiation that the Sun produces. For example, every four square meters of horizontal ground surface in the Sahara region receives the thermal equivalent of one horsepower from solar radiation every year. Consequently, no less than 25,000 horsepower years can be obtained from one square kilometer of the Sahara's surface, assuming efficiency of only ten percent. On page 11 we quantified the Earth's probable energy consumption in the year 1970 at a billion horsepower years. This energy demand could be met by 40,000 square kilometers of the Sahara's surface, while the entire Sahara is 150 times bigger than that, even if only ten percent of the solar heat is utilized. Thus, if solar heat can be made usable, then humanity's energy problem is solved with a single blow.

There have, in fact, been attempts to achieve this goal. So far, none of them has had a useable result, since they were all undertaken with unsuitable means. It is pointless to build giant mirrors and complicated evaporators to achieve a final output of 50 horsepower. We must free ourselves of the traditional view that the intermediary between heat and electrical energy must, under all circumstances, be the boiler. The boiler is useful when the heat is available in concentrated form, as in the case of coal, however not when it is "diluted", distributed over country-sized areas, as is the case with solar radiation. In order to make such heat usable, we must use Nature as a model. Gandrillon showed o n e way to do this, which, however, has limited usefulness (pp. 20 ff.). Bernard Dubos pointed out a completely different plan that even the Paris Academy, which is not exactly progressive, found so exciting that it recommended it for North Africa, which lacks coal. This plan can be summarized as follows in o n e sentence: The right means to exploit solar heat is the wind! For anyone who knows the many unsuccessful attempts to make use of air currents, this sounds paradoxical. The wind's intermittent nature seems to make it the worst conceivable basis for large-scale power supply. However, it is not natural air currents that Dubos has in mind. While Gandrillon wants to use the Sun to create artificial waterfalls, Dubos thinks of using it to produce vertical air currents like cyclones. Let us listen to what the inventor has to say about it.

If we could, for a moment, convert the air into water to make it visible, then everyone would see what meteorologists and pilots have already known for a long time: that a good part of the sea of air around us consists of currents, whirlpools, and vortexes. And, in this world which is normally invisible to us, we could hardly take a step without everywhere encountering "rising waterfalls", such as are seen in films showing a waterfall flowing backwards. Since so many actual waterfalls are exploited now, it is entirely logical to think of exploiting rising flows of air as well. After all, tropical cyclones show us how much energy an air current can hold. And a moment's reflection tells us that these currents are just as inexhaustible as those of water, since they are both produced by the same source: the Sun. Meteorologists have investigated these things for a long time. But their knowledge has remained incomplete, since no one has united the individual facts into the big picture. I have a textbook of meteorology in front of me. In it, I read that the atmosphere is constantly circulating in flows that are just as unchanging as the Gulf Stream and whose paths are just as well known. The best example is provided by the trade winds, which always move toward the equator, where they create a powerful, almost vertically rising air current. Their counterpart is at the poles, where the cycle is closed by "falling winds". – I page around through the book and read that if the barometer in the area of a cyclone falls from 760 millimeters – which is the normal pressure – to 715 millimeters, this difference in pressure produces wind speeds of up to 180 kilometers per hour. Such storms have violent energy, whose devastating effects are known. Another place in the book says that the yearly average air pressure, which is 760 millimeters at sea level, is only 603 millimeters at an elevation of 1,000 meters; thus, the

difference is much greater than the pressure
differential in the area of a cyclone. The air
temperature also decreases quickly with increasing
altitude, as is shown for the open atmosphere by the
curve in Figure 11.

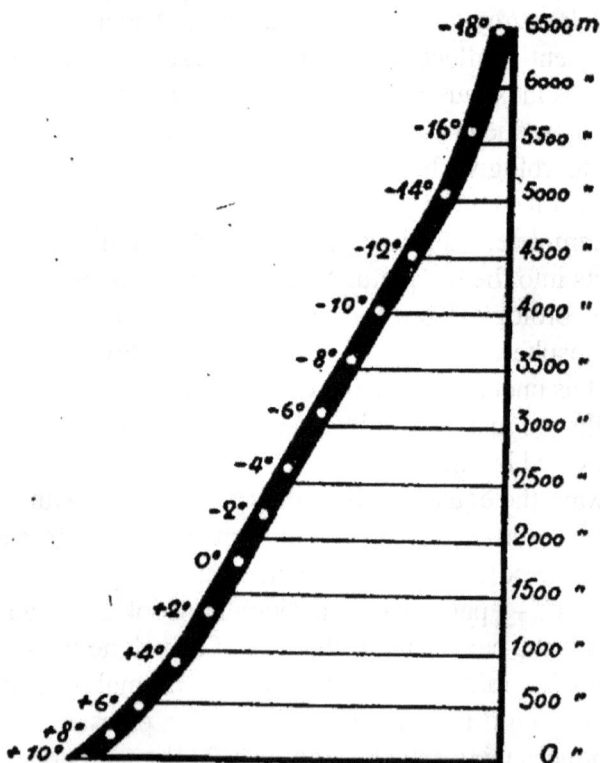

Figure 11. Decrease of temperature in the open
atmosphere with increasing altitude. This is one of
the facts that Dubos makes use of in his wind power
stations.

Dubos combined these observations and, on the basis of the resulting facts, put together a project that directly connected the hot, high-pressure air of sunlit plains with the low-pressure zone on the mountain peaks. In this way, the inventor wants to unleash artificial cyclones. The only difference is that their energy is not supposed to cause devastation when released, but instead be well-behaved and industrious and drive turbines, like their sibling the waterfall. The significance is obvious: this is just what we want. The design is also simple, as is the case with all truly ingenious ideas. To accomplish this, a pipe that is about 1,000 meters long and open on both ends is leaned as steeply as possible against a mountainside in a quite hot area – the Sahara for instance (see Figure 12). At the bottom end, the pipe widens into an enormous funnel, which for all practical purposes has the shape of a large, flat greenhouse and which, when heated by solar radiation, acts like a powerful sauna and increases the air temperature to a maximum value. The fact that the greenhouse opens wide in front but narrows in the shape of a funnel toward the back simultaneously has another result: The inflowing air enters into a vortex motion, exactly the same as at the beginning of a cyclone. Thus, the air enters into the pipe, which one should imagine to have a very large diameter and very smooth inside walls, in order not to slow down the airstream. However, the outside of the pipe is surrounded by thick heat insulation, so that the warm air loses as little heat energy as possible as it rises. The reason why is that the greater the temperature difference between the inside of the pipe and the outside air up at the peak, the greater the speed at which the airstream rushes through the pipe. According to calculations, it should reach 180 to 200 kilometers per hour. In practical terms, such an airstream is quite comparable to a cyclone which, when harnessed, can only unleash its fury in one place: by throwing itself against the rotor blades of the powerful air turbine at the top of the pipe. Of course this turbine, like its sister in the water stream, is coupled with a generator that produces electric current. In this form, the energy of the swirling airstream flows back into the lowlands, where it powers other machines.

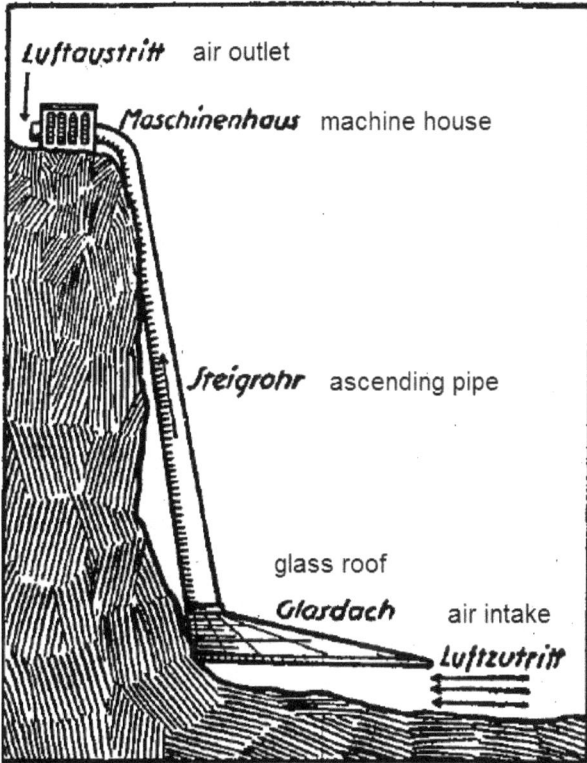

Figure 12. Diagram of the Dubos wind power station: A sunlit plain at sea level is connected with the low pressure zone on mountain peaks by a 1,000 meter high ascending pipe. The result in the ascending pipe is a constant rising airstream having a speed of perhaps 50 meters per second, whose immense energy can be exploited in wind turbines. The glass roof at the foot of the ascending pipe produces superheating, which further increases the speed of the airstream.

This idea was a clever deduction drawn from laws that have been confirmed hundreds of times by precise observations but that have not had any productive application so far. Anyone can test the soundness of this idea with a simple experiment. First, take a piece of sheet metal; this represents the plain. Beneath the sheet metal, place an alcohol flame; this represents the Sun's thermal radiation. The supplied heat raises the sheet metal's temperature somewhat above that of the surrounding air. As a result, a warm air current rises from its surface; this current can clearly be felt by holding one's hand or face over it. Now we take a miniature turbine, a simple pinwheel like the ones sold at amusement parks, and fasten it about 25 centimeters over our miniature plain, horizontally of course, so that the airstream strikes the vanes the right way (see Figure 13). What happens? Nothing! The air current is too weak to turn the pinwheel; it does not move even if the alcohol flame is made larger to increase the supply of heat. However, if we take a tube that widens into a shallow funnel at the bottom and insert it between the turbine and the sheet metal shown in Figure 13, then the wheel immediately begins to rotate. The speed increases quickly and shows the presence of a powerful airstream. If we pull the funnel tube out, then the wheel stands still. If we put the tube back in its place, then the turbine runs again. As simple as the experiment is, it is conclusive. There is no doubt that we are dealing with a quite natural process. That is, we may also assume that the enormous quantities of warm air that constantly form over a real plain whose entire surface is heated by the Sun can be converted by enormous funnel tubes into powerful rising air currents, and in this way be utilized for energy production.

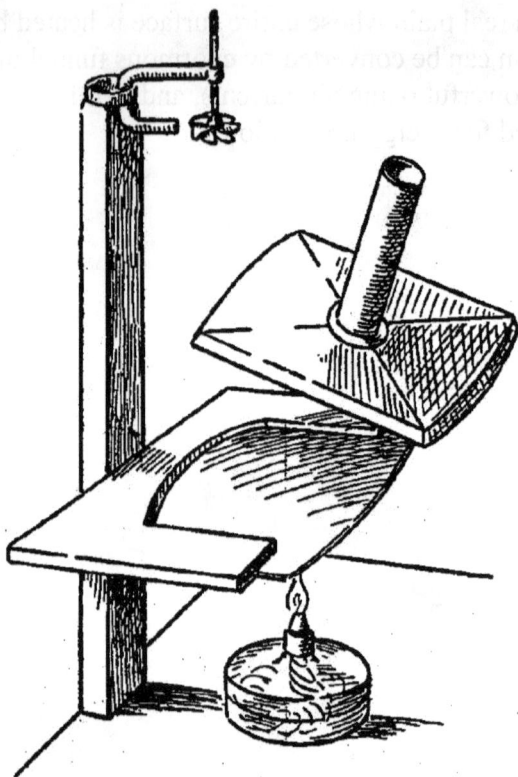

Fig. 13. A simple experiment that Dubos used to prove the correctness of his ideas. The sheet metal represents the Sahara, and the alcohol lamp represents solar heat. The small pinwheel at the top is the wind turbine. If the alcohol flame is pushed under the sheet metal, hot air currents rise. However, the turbine only works once a funnel placed onto the sheet metal forces the air currents into a certain path, concentrating them in this way.

After this confirmation of theory by practice, the question still remains whether technology is capable yet of realizing plants like those that Dubos wants to make. A glance at Figure 12 is sufficient to answer this question in the affirmative.[13] The greenhouse does not present any difficulties, and a few hydroelectric power stations have long had tubes of similar length; the Pougny power station in the Rhône valley even uses a drop of 1,630 meters to drive its turbines. In order to withstand the pressure of the falling water, such pipelines must be made of seamless drawn steel pipes that must be carefully anchored in the ground. Here, however, where we are dealing with more of a chimney, it is easily possible to manage with much lighter constructions, for instance with a reinforced concrete tube 10 meters in diameter and 1,000 to 1,500 meters long. The problem of maintaining the air's temperature as much as possible as it rises is also not difficult to solve, since there are several suitable ways of doing this. It is perhaps most expedient to make the tube out of lightweight concrete, whose innumerable air voids are excellent thermal insulation. Thus, this only leaves the large wind turbine that still has to be built. This task will certainly not be too challenging for our engineers, with their experience in hydroelectric and steam power.

---

[13] Translator's note: In fact a solar chimney (solar updraft tower) really was built in Manzanares, Spain in 1981. It was destroyed in a storm in 1989.

From all this, it also turns out that the construction costs of such power stations cannot be large. In no case do they exceed the costs of large hydroelectric power stations, and they are probably even less, since here there are no dams, surge tanks, weirs, grills, channels, etc. It is obvious that only hot countries are suitable for such plants. Dubos says that that the best regions for this are those between the isotherms for + 20°C above and below the thermal equator on the schematic map in Figure 14. The significance is obvious: This is an immense area, but it covers almost only very sparsely populated countries, including, among others, all of Africa. Here above all the Sahara offers favorable conditions, since three chains of the Atlas Mountains and Hoggar Mountains with their steep drops to the desert are almost tailor made for assembling ascending pipes, while Algiers, Tunis, and Morocco have long been desperate for new energy sources. Therefore, it is probably here that the first experiments will be done. If they are successful, in a hundred years the mountain slopes of North Africa and other hot countries will probably appear as they do in Figure 15: full of countless wind power stations whose turbines not only supply the surrounding areas with current, but are also connected with the large power networks of Europe by intercontinental high-voltage lines. This would make it possible to exchange current and thus eliminate the single difficulty that such wind power stations obviously present: that they can only be operated during the daytime. During the daytime, regular operation is quite definite, since the airstream along the floor of the ascending pipes is constant and always of the same size. It is important to understand that the immense quantities of cold air up in the atmosphere over the "catchment areas" of such wind power stations form a fundamental part of the system, exactly like the quantities of water over a waterfall belong to the system of the power station fed by this waterfall. These enormous quantities of air will constantly maintain the air currents as long as the ground is warmer than the atmosphere. In these areas this is always the case during the daytime, but at night it is surely doubtful, in view of the rapid cooling after sunset. For this case, an integrated economy makes it possible to even things out. During the daytime, current is supplied from Africa to Europe into the thermal and hydroelectric power networks there with their high daytime current demands. At night, current is supplied by European plants, which have enough nighttime power available, into the African wind power networks, whose machines could then be idle. Thus, this project, which opens up to us an energy source that is enormous from our perspective, also encourages the very close connection of Europe with Africa that is emphasized by the Panropa plan. For the technology of the future that thinks in terms of continents, political borders have no meaning.

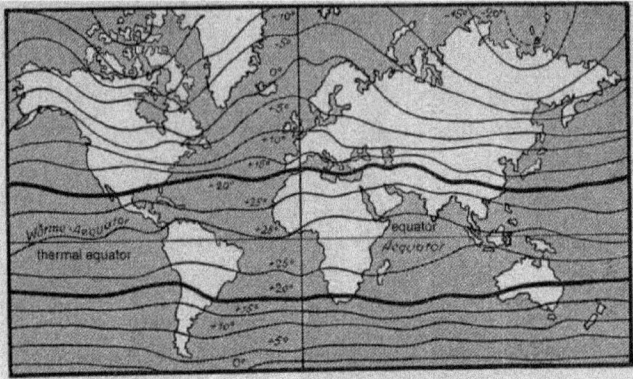

Abb. 14. Erdkarte mit Wärmelinien, die die Temperaturverteilung (Gegenden gleicher Temperatur) sichtbar machen

Figure 14. World map with isotherms showing the temperature distribution (areas of equal temperature).

Figure 15. A picture of the future: Wind power stations at the steep drops of the Atlas Mountain chains in the Sahara

6. Wind Towers

Since the wind power stations of Dubos are tied to
those areas where solar radiation reaches its
maximum value, in Europe they can only be used in
the far south. Thus, if we want to use wind power for
energy generation in the colder countries as well, we
must look around for other methods. On a small
scale, this has already been done for a long time,
right in Northern Europe, since The Netherlands and
Northern Germany can surely be called the home of
the windmill. Everyone knows that the old windmill
is not a useable "wind machine". The same goes for
its successor the wind turbine, which can also be
considered only as a small power machine. There
have been many attempts to improve its limitations.
The utilization of wind power has been a favorite
problem of inventors since time immemorial. All
existing designs up to now are based on the same
principle: the wind drives a propeller with multiple
blades, a process that is a mirror image of how the
fan works. In theory, the process is clear and
irreproachable; however, putting it into practice
presents all kinds of difficulties. First it is necessary
to arrange the propeller at the top of a tall frame
structure outside, where it is exposed to the rigors of
the weather. Then, whenever the wind changes
direction, the heavy wind wheel must be
correspondingly turned, which unfortunately still
does not prevent the efficiency dropping to almost

zero if the wind's attack angle is unfavorable. Furthermore, it is necessary to transfer the rotation of the propeller shaft down to the ground by a linkage, resulting in a large loss of energy. Finally, if a wind wheel is supposed to deliver rather significant power at an average wind speed of five meters per second, it must be extraordinarily large, since the energy available in natural wind decreases very quickly as the wind speed goes down. All this clearly shows that it will never be possible to create true wind p o w e r   p l a n t s on this basis.[14] If this is desired, then other approaches must be sought that have not yet been tried. Such approaches have been developed through research conducted at the Institut AéroTechnique in Saint-Cyr, France, for entirely other purposes: to create a powerful fan, working exclusively with natural wind energy, to ventilate living and working areas, and to increase chimney draft. The task was solved, and had the unexpected result that during testing the experimental plant simultaneously turned out to be an ideal wind power machine having an efficiency of almost 100%. Figure 16 clearly shows the system, which has no moving parts on the outside. On the outside, all one can see is a cylindrical tube resembling a tower, a type of chimney made of sheet iron, whose top is encircled at a certain distance by a rather short cylinder. This ring is held at the bottom by narrow

---

[14] Translator's note: As we know, wind turbines have now been built that are large enough to make wind power plants possible. However, the discussion that follows is still relevant.

supports, but in other respects it is open on both sides. Everything else follows from Figure 16. No matter what direction the wind is coming from or what its angle of attack, the result is always the same: a vertical airstream in the cylindrical tower. In other words, here the unsteady outside wind, whose strength and direction are variable, is literally converted into a steady airstream that always flows in the same direction, although its speed obviously varies. Thus, it is sufficient to arrange a horizontal turbine at the lower end of the wind tower to have power that is proportional to the wind energy and that only stops when the wind does, i.e., never. It is this arrangement of all moving parts – the turbine and the generator driven by it – on the floor of the wind tower that presents the second greatest advantage of such wind power plants. This makes the machines easily accessible, and simultaneously protects them from all the rigors of the weather, exactly like in a conventional electric power plant. Finally, a third advantage is that the efficiency is almost 100 percent, since the speed of the air drawn in is always equal to the wind speed, so that the only source of losses that has to be taken into account is friction. This principle makes it entirely possible for wind power plants to have ever larger capacity. Figure 17 shows how they will look, a fantasy whose realization we all may live to see, since already today the wind towers could be valuable sources of power in areas not yet covered by the public electrical power supply. However, at a later point in the future, hundreds of them will be combined into batteries on plateaus and mountain ranges, thus creating true large power plants to exploit wind energy, which might be the only possible way, aside from hydroelectric power plants, for Central Europe to be self-sufficient in energy even after exhaustion of the country's coal deposits.

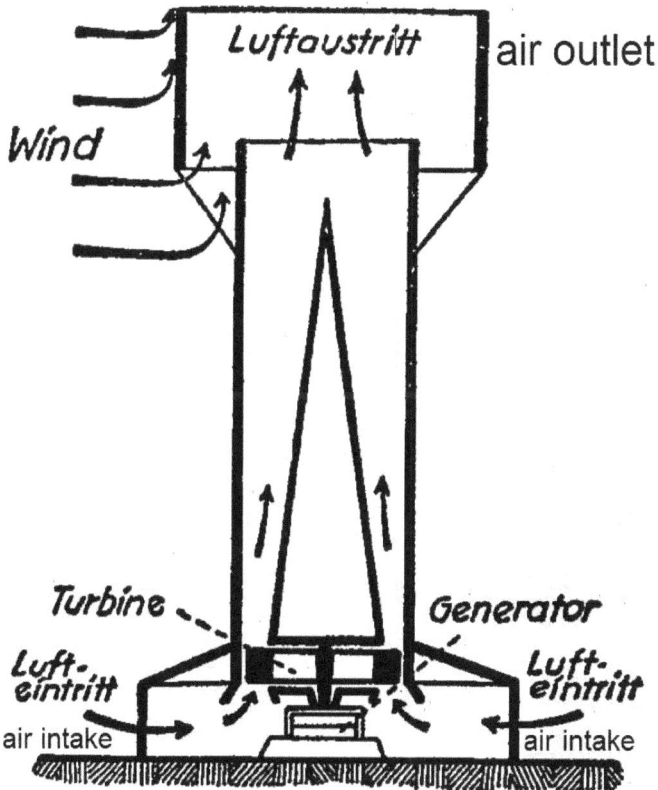

Figure 16. Schematic representation of a wind tower

Figure 17. Wind towers; another proposed method of exploiting wind energy

7. Tropical Oceans as a Heat Source

15

On September 17, 1881, the "Revue Scientifique"
ran an extremely strange article whose author was
Prof. d'Arsonval, the physicist who later became so
famous. "Imagine that we immerse a boiler into the
30°C water of the artesian well of Grenelle", he
wrote in this article. "If we then feed tap water that
has an average temperature of 15°C into the
associated condenser, we obtain a temperature drop
of 15°C. Imagine that the boiler is filled not with
water, but with liquid sulfurous acid, which has a
vapor pressure of 343 mm Hg when heated to 30°C.
In the condenser, the pressure would then be 206.5
millimeters. This pressure difference corresponds to

---

[15] Author's note: The second law of thermodynamics states that
to convert heat into mechanical energy it is always
necessary to have t w o temperature levels, i.e., a
temperature difference. For example, in the gasoline engine
we have the combustion temperature of the gasoline vapor
behind the piston, on the one hand, and on the other hand
we have the cold surroundings, while in the steam power
plant the two temperature levels are those of the hot
combustion gases of the coal under the steam-producing
boiler and the coolant of the condenser that recondenses this
steam. Conversely, it follows from this law that every
temperature drop is an energy source. Knowledge of these
facts is a prerequisite for understanding the following
discussion.

a constantly usable pressure of almost two atmospheres, which does not cost us anything at all.

"Does Nature offer such conditions to us frequently?" the article goes on to ask. "Of course", is the answer, "since there is no absence of hot springs; however, in addition there are also other usable heat sources here. The reason why is that we could just as well set our condenser in glacial ice and the boiler in 15°C river water: in this case we also have a temperature drop of 15°C, and thus the same vapor pressure as before. – The ideal would be to immerse the boiler in the sea near the equator and the condenser into the eternal ice of the poles. However, in practical terms it is not at all necessary to build such a long pipeline. The reason why is that the deep water of the sea has a temperature of only 4°C everywhere – even at the equator. Thus, it is sufficient to put the boiler on the water's surface and the condenser 1,000 meters below it, in order to have a sufficient temperature drop."

That is how far d'Arsonval went! We can imagine how half a century ago people smiled at these ideas of a young physicist who was almost unknown at that time. His ideas probably were not even considered worth discussing. The reason why we think this is that even later inventors had the same experience when they presented the same idea to the public and explained the theoretical foundations much more carefully: the American Campbell (1913), the Italians Dornig and Boggia, and a

German physicist, Dr. E. Bräuer, who might be the one who elaborated the problem most thoroughly. No one listened to any of them, and what they wrote gathered dust, just as d'Arsonval's article had. The world did not take notice until November 1926, when the Paris Académie des sciences received a report whose author was world-famous.

"I have the honor", began the author, "in the name of P. Boucherot and myself, to present the Academy with the results of studies that allowed us to learn totally unexpected facts in the field of physical geography.

"As is known, at depths of 1,000 meters sea water has a uniform temperature of 4-5°C year in, year out. On the other hand, it is known that the surface temperature of the tropical seas generally lies between 26°C and 30°C. Starting these two facts, it is possible to devise an ambitious plan to exploit the heat of the sea, which is ultimately a plan to exploit solar heat.

"At first, making use of the deep water appears to present difficulties. However, they are easy to overcome, since it is sufficient to submerge a tube that has good thermal insulation down to the desired depth. Then, the cold water rises in this tube. Since the specific gravity of the deep water differs somewhat from that of the surface water, the deep water will not quite reach the top edge of the tube, but rather its level will be about a meter lower. Thus, raising the water only requires pumping over this height; it rises the rest of the way according to the law communicating vessels.

"At first, the fact that the temperature difference between the deep and the surface water is only 20-22°C appears to present an obstacle that is difficult to overcome. The resulting vapor pressure is so small that it has never been utilized up to now as a source of energy, because it has been assumed that it would not suffice for this purpose. However, our experiments have quite unexpectedly taught us differently.

"Mostly out of curiosity, we checked whether the vapor pressure in question was sufficiently large to exploit as a source of energy; we did not have special hopes that it would. However, to our greatest astonishment it turned out that the energy of that water vapor really was sufficient for current steam turbines, although the pressure differential lies far below the usual values. If water vapor of 24°C, which has a pressure of 0.03 atmospheres, is drawn off by a high vacuum created by cooling to 7°C in the condenser, then the vapor has a flow rate of 500 meters a second and drives a single-stage turbine at a rotational speed of up to 250 meters a second. Each kilogram of this vapor, whose pressure is 700 times smaller than that of a vapor of 20 atmospheres, performs work that is only five times less than the quantities of energy that are obtained by expanding vapor from 20 atmospheres to 0.2 atmospheres. This fact was so surprising that we devised a small experiment to make it believable. We took the rotor of a de Laval turbine built for a pressure of 20 atmospheres and installed it horizontally in a vessel as is shown in the attached sketch (see Figure 18). A feed pipe conveyed 28°C water vapor from a large bottle down onto the turbine wheel. Broken ice was laid in the space below the turbine; it acted as a condenser. At the top of the turbine vessel there was a tube through which an air pump slowly sucked the air out of the two containers. As soon as this caused the pressure inside the water bottle to fall below the water's vapor pressure, the water began to boil [16]; the vapor flowed through the turbine, started it rotating, and recondensed to water under the influence of the ice. The turbine, which turned at 5,000 revolutions per minute, was used to drive a small generator, which delivered enough current to make three small electric light bulbs shine brightly.

---

[16] Author's note: This happened according to the known fundamental principle of physics: the lower the pressure, the lower the evaporation temperature. In the mountains water boils more easily than on the plains, since the air pressure is much lower up there. If the pressure is lowered enough, then water already boils at room temperature.

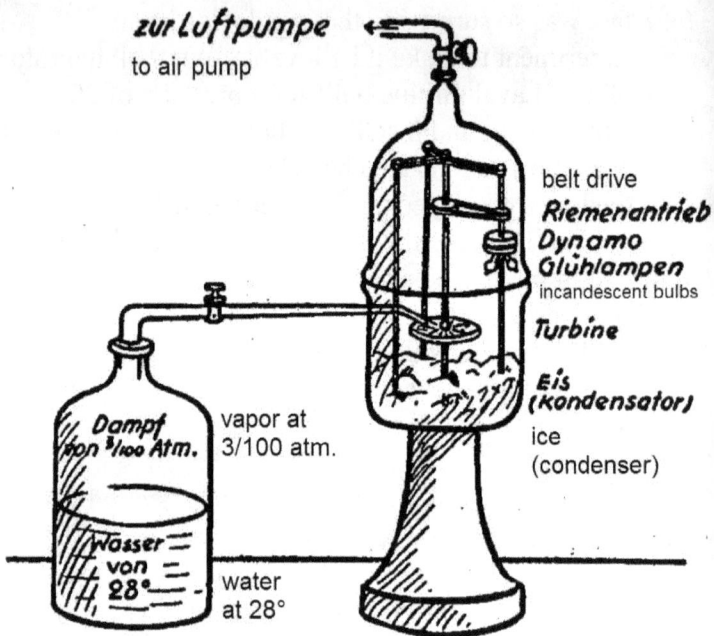

Figure 18. The experimental apparatus that Claude and Boucherot used to demonstrate their process for generating energy from the sea before the Paris Académie des sciences.

"This experiment showed, on the smallest scale, what we want to carry out on a large scale. The artificially heated water is replaced by the sea's warm surface water. It is continually drawn into a boiler in which it evaporates under reduced pressure. After the water vapor has passed through the turbine, it is recondensed to water by cold water in the condenser. The vapor pressure is only 0.03 atmospheres, i.e., the flow that is continually going through the apparatus is hardly noticeable. However, if it should be possible to build a turbine that can be operated with such low pressure, then nothing stands in the way of economic exploitation of the process, since the necessary warm water is always available in unlimited quantities."

The project's author was Georg Claude. This ensured that the short report would attract the greatest attention everywhere. Otherwise, the idea would probably also have been dismissed once again as the product of the mind of a dreamer. Experts have long known that Claude is no dreamer. They also know that it has always been his principle to leave the beaten paths to search for the unknown on new paths. This has already brought him great industrial success several times, sometimes in ways that others would have declared impracticable. From the start, this fact ensured that he was received quite differently from how a theorist would have been, no matter what his reputation. Nevertheless, more conservative experts greeted the idea with extreme skepticism. Some found fault with this detail, others with that one. Yet others tried to prove with numbers that the entire proposal was nonsense. To all these objections Claude remained silent. Instead, he did the cleverest thing that he could have: He decided to answer the armchair critics with the practical results of public experiments, whose scale was gradually increased.

First, the experiment described above (see Figure 18) was repeated before the members of the Paris

Académie des sciences and a number of invited
guests, and this experiment was completely
successful: the turbine ran, the electric light bulbs
burned, and the power output was measured at 3
Watts. The second experiment was of a larger scale.
Claude and his collaborator, the engineer Paul
Boucherot, designed a 50 kilowatt system, which
was put into operation at the Ougrée-Marihaye iron
and steel works in Belgium. The source of heat was
the coolant from the blast furnaces, whose normal
temperature of 14°C was increased enough by the
introduction of steam that it was always 20°C
warmer than that of the water from the Meuse that
was used to cool the condenser. Thus, for the entire
duration of the experiment it was possible to
maintain the same temperature drop that would,
according to all available experience, be expected in
the tropical seas. Pumps and deaerators were
installed on a corresponding scale. On April 25,
1928, the system was ready for operation. Four days
later the turbine was running for the first time, at
5,000 revolutions per minute and a calculated output
of 50 kilowatts right from the start. A check of the
energy balance showed that all the auxiliary
machines consumed only ¼ of the power that was
generated, so that the useful yield was 75 percent.
This proved that Claude's ideas were useable in
practice.

On June 1, 1928, the pilot plant was visited by a
commission of the Paris Académie des sciences. Its
reporter, Le Chatelier, concluded his expert opinion

with the words: "This is the first time that anyone has ever seen a steam engine operating at a temperature difference of a few tens of degrees. This appears to solve the first part of the problem; the second remains, that of lowering a several thousand meter-long pump pipe into the sea."

This statement expressed doubt and issued a challenge. Claude did not hesitate to meet it, and did so by investing his entire fortune, since the costs of the large-scale experiment that now followed amounted to several million marks. The site that was selected for this experiment after careful pilot studies was the Bay of Matanzas on the north coast of Cuba, about 85 kilometers east of Havana. There the water temperatures, the slope of the ocean floor and other conditions are especially favorable. This third system is shown in Figure 19. At W, pump A sucks in 28°C surface water and pushes it into evaporator B after it has been deaerated. Here the evaporation pressure is lowered far enough by creation of a vacuum that water's boiling point is reduced to 28°C. Consequently, the water that is fed in evaporates without further addition of heat. The vapor goes through pipe C into low-pressure turbine D, which is coupled with electric generator E, causing it to rotate, and then the vapor goes on to condenser F. This is where the cold deep water K comes into play. Pump G sucks in cold deep water K at the bottom of the sea through pipe H, which brings it to the surface, and the cold deep water is pressed into the condenser, where it recondenses the exhaust vapor of the turbine back to water. The condensed water, along with the coolant, flows back into the sea through pipe L. Condensation of the exhaust vapor creates a vacuum in the condenser whose negative pressure sucks in the vapor driving the turbine. As soon as this happens, the vacuum pump used for startup, which evacuates the air from the evaporator, can be turned off, since from that point on the heat drop between the surface and the deep water keeps the system in operation, as long as pumps A and G are running.

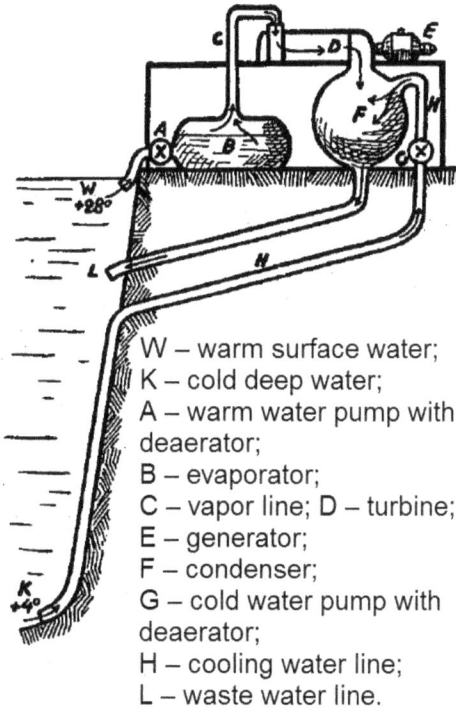

W – warm surface water;
K – cold deep water;
A – warm water pump with deaerator;
B – evaporator;
C – vapor line; D – turbine;
E – generator;
F – condenser;
G – cold water pump with deaerator;
H – cooling water line;
L – waste water line.

Figure 19. Diagram of the pilot plant that Claude and Boucherot set up on the north coast of Cuba

This short description clearly shows that, disregarding the size of the individual parts, the difficulties involved in constructing such a plant lie at essentially one place: laying the gargantuan suction pipe that brings the cold deep water to the surface. For the Matanzas site it turned out that the pipeline had to be around 2 kilometers long. The reason why was that there the pipe could not be put down vertically, but rather had to be laid at a slope on the gently descending ocean floor. Since Claude was thinking of utilizing the pipe for a larger plant later, his plan called for an inside diameter of 1.6 meters, although a much smaller diameter would have sufficed for the pilot plant. It was made of slightly corrugated sheet steel, first in lengths of 20 meters, which was given the necessary thermal insulation and then was bolted together using rubber gaskets. Two parts were manufactured in this way, one piece 150 meters long, that was pushed out from the coast into the sea, where it reached down to a depth of 18 meters, and the main pipe, which was 1,850 meters long and which divers were supposed to connect to the coastal pipe. The pipe laying was unsuccessful two times. The first time the cables holding the pipe broke, and it sank; the next time it bent, which put a crack in it that could not be repaired. These accidents meant huge losses of money. Nevertheless, Claude did not lose his courage for a moment. On September 7, 1930, a third pipe was laid, and this time everything went according to plan. However, the new pipe was shorter than the first, since it already ended at a depth of 600 meters in the sea.

The next weeks were spent assembling the auxiliary machines, and on the first of October the turbine ran for the first time. However, it produced only 22 kilowatts. The reasons were obvious. The suction pipe was too short to bring up the deep, cold water. Consequently, the cooling water coming into the condenser was at about 14°C, so that if the surface temperature was 28°C, a heat drop of only 14°C was available. So then Claude installed, in Matanzas, the small 50 kilowatt system originally used in Belgium, for which the new suction pipe, with its centrifugal pump dimensioned to convey 4,000 cubic meters of cooling water an hour was much too big. Under these circumstances, the auxiliary machines consumed considerably more energy than the turbine output.

For the superficial observer, this proved the failure of the entire principle. By contrast, if we get to the bottom of the matter, then we have to agree with Claude in concluding that in any case the experiment proved one thing: in the tropics it is theoretically possible to exploit the heat stratification of the sea to generate energy. The fact that this presents difficulties and that therefore the first attempts did not meet the expectations that were hoped for is not a reason to give up. There is not a single earth-shattering invention whose realization would not require some groping and searching, even if its creators were completely certain about its theoretical foundations. Even Claude did not let himself get at all discouraged by the outlined result; on the contrary, instead he attached the boldest hopes to the apparent failure. The following considerations led him to this: In a large power plant, a suction pipe whose dimensions equal those of a railway tunnel will suck water up from the bottom of the ocean, where its temperature is only about 4°C. Then a heat drop of up to 24°C can be expected. Since the thermodynamically obtainable power is proportional to the square of this temperature difference; since the efficiency of the steam turbines increases quite significantly with their size; and finally since the frictional losses – and thus the work that the cold water pump expends for every cubic meter of water it conveys – decrease rapidly as the diameter of the supply pipe increases, the efficiency of a future large power station must be much more favorable than that of the experimental power plant at Matanzas.

Using the results of the previous experiment, Claude calculated the basic factors of such a future plant. He came to the conclusion that after all auxiliary work is subtracted, every cubic meter of water conveyed from the depths will perform net work of 50,000 kilogram meters that is available in the form of electrical energy. In other words, the plant's power and water consumption will be the same as that of a conventional hydroelectric power station with a drop of 60 meters. Such a plant of Claude's design would require construction costs of 400 marks for the installed horsepower, while 500 to 800 marks are expected for large hydroelectric power stations. An ocean power plant need not acquire any water rights, its upkeep is considerably less expensive than that of a hydroelectric power station, and finally sea water is constantly available in unlimited quantities, so anyone can see that Claude's ideas definitely have a future, even if the inventor, in his optimism, underestimated the real production costs. Finally, it must also be emphasized that when calculating the profitability of such plants, it is a mistake to base the calculation only the conditions prevailing today. After all, these are new ways of doing things, and today's economic misgivings will probably be outdated in a few years. In any case, the history of engineering offers an abundance of examples of how easy it is to make a fool of oneself to future generations by saying "impossible". Our innate mental inertia makes us only too ready to expect everything in existence to have a constancy which does not at all exist in reality. For example, let us assume that the idea, which is already frequently articulated, of anchoring floating airports as bases on the southern air route over the Atlantic should take shape. Claude plants would be the best way to generate the operating current for the machines, floodlights, etc. Of course in such cases, the suction pipe would be lowered vertically, a laying method that is technically much simpler and considerably less expensive than laying it on a slope from the coast. The reason why is that the pipe could then be made light so that its water displacement was equal to its weight, so as not to produce any tension in its wall. Incidentally, it also seems that Claude ultimately had similar ideas. This is suggested by the embodiment of a Claude power plant shown in Figure 20, in which the entire plant is put on a giant floating pontoon. – However that may be, it would be a gross underestimation of a researcher having the stature of Claude and a very experienced engineer like Boucherot at his side, to dismiss the whole question on the basis of short-sighted calculations that both men certainly also carried out themselves. Therefore, "Mechanical Engineering", the great American engineering magazine, was quite correct when it wrote in December 1930: If Claude and Boucherot nevertheless decided to carry out an experiment that costed considerably more than one million dollars, which they did not raise by selling stock but rather took from their own funds, then experts should refrain from expressing excessive doubts about the venture; this is because it is precisely the gift of seeing a possibility in a place where no one else does that distinguishes the successful inventor from all those who use calculations to predict the impossibility of success.

91

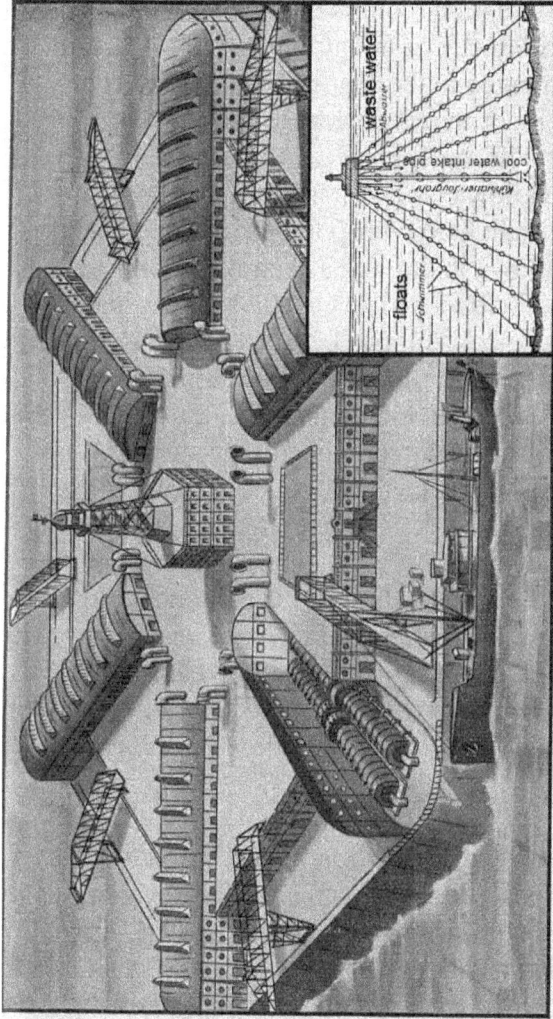

Figure 20. Claude and Boucherot's plan for a floating sea power plant. (cont'd. next page).

Figure 20, caption (continued): A hexagonal pontoon having a diameter of 600 m supports six generator rooms arranged in the shape of a star. Each room contains four turbine generators each having power of 40,000 kilowatts; that is, the total power is around 1 million kilowatts. The warm surface water is drawn in right below the pontoon; the cold deep water is drawn in through a central suction pipe that is 600-1,000 m long. The generator rooms have electrochemical factories between them that convert the generated current into easily transportable chemical products. Loading is done through movable conveyor bridges directly into ocean steamers. The central block contains the living space of the engineers and workers. The lighthouse on top of the central block shows the position of the "island". Such pontoons could already be built today using modern means of shipbuilding. Anchoring is a problem in and of itself. The inset sketch shows how this is envisioned.

*   *   *

However, if Claude's ideas should, contrary to expectations, prove to be unworkable, there is still another way to reach the desired goal. This was already pointed out by Wilhelm Schmidt, the famous inventor of the superheated steam engine, in the fall of 1922 and, independently of him, by the Berlin physicist Dr. E. Bräuer in July 1924 – that is also even before Claude did. The main difficulties in Claude's idea obviously have to do with the necessity of degassing the surface water before evaporation [17] and with the colossal dimensions that the entire engine plant requires to produce really significant power, in view of the low vapor pressure. Of course the reason for the colossal dimensions is that the lower the vapor pressure, the greater the quantity of vapor required to operate a turbine, which in turn requires corresponding quantities of warm water and cooling water. All these difficulties disappear if the surface water is not directly used to drive the turbine, but rather used to heat a water-tube boiler. In this boiler, some low-boiling carrier fluid, e.g., ammonia or liquid carbon dioxide or sulfurous acid, as proposed by d'Arsonval, is evaporated under relatively high pressure. This high-pressure vapor then drives the turbine. Afterwards, it is condensed in a condenser cooled by deep water, and fed back to the boiler in this form [18]. The advantages of this system, developed mainly by Bräuer, are obvious. It is unnecessary to degas the surface water. It is also unnecessary to evacuate all the air from the evaporator to lower the evaporation temperature. This eliminates two auxiliary machines that are indispensable in Claude's system, and these two

---

[17] Author's note: This is because it contains a considerable amount of air that would be released when the water is evaporated, and its partial pressure would reduce or even eliminate the very small pressure difference between the evaporator and the condenser.

[18] Author's note: In theory, such a system completely corresponds to that in Figure 25. The pipe carrying the natural steam in that system would carry surface water in this system.

auxiliary machines consume a good portion of the power produced. However another fact is of even greater importance. In a water-tube boiler that has warm water flowing around it, the heat transfer at the tube walls is several hundred times more favorable than it is in a similar coal-heated boiler. Consequently, the boiler in this binary system is not significantly larger than that in a conventional steam power plant, despite the small heat drop that is available. The same goes for the turbine, which is, after all, operated here with a much higher vapor pressure, so that it can be relatively small. Finally, Bräuer does not want the cold water pipe to go anywhere near as deep as Claude did. The reason why is that according to his calculation, which has been checked and found correct by different parties, a binary system of the type being discussed works most economically at a temperature drop of 20°C. In other words, if we start with surface water at 28°C, the cooling water can be 8°C. However, in many places in the tropical seas water of this temperature is already found at a depth of no more than 400 m. In economic terms, these changes have the effect of a substantial cost reduction. This applies not only for the plant itself, but rather also for the technical expense involved in assembling and operating it. Bräuer assumes, on the basis of careful calculations, that the costs of the plant will not be any higher than those of other thermal power plants, while the operating costs will be substantially lower. For example, for a power plant having a power of 2 million kilowatts he estimates the per-kilowatt plant costs to be 300 marks.

It must be admitted that this train of thought is convincing. But here again it is possible to raise the objection that we already expressed above: Claude and Bräuer both know all these things. Thus, if he nevertheless stuck to his plans and risked millions of his own money for them, he must certainly have good reasons for doing this. The explanation might possibly have to do with a circumstance that we have not yet mentioned: Since Claude's system evaporates the seawater itself, it will deposit large quantities of salt in his evaporators, and this salt will contain all kinds of valuable substances. The necessity of removing these quantities of salt is another objection that critics have made against Claude's system. But what if Claude sees this as the best economic prospect for his system? From the heat engineering perspective, the presence of the salts is not, of course, desired. However, in economic terms under some circumstances these "by-products" could be much more profitable than the energy generated. It is easy to imagine the engine plant, whose general arrangement is clearly shown in Figure 21, being designed in such a way that only one part of the evaporator works at a time, in alternation. Then, cleaning out the accumulated salts causes no interruption of operation.

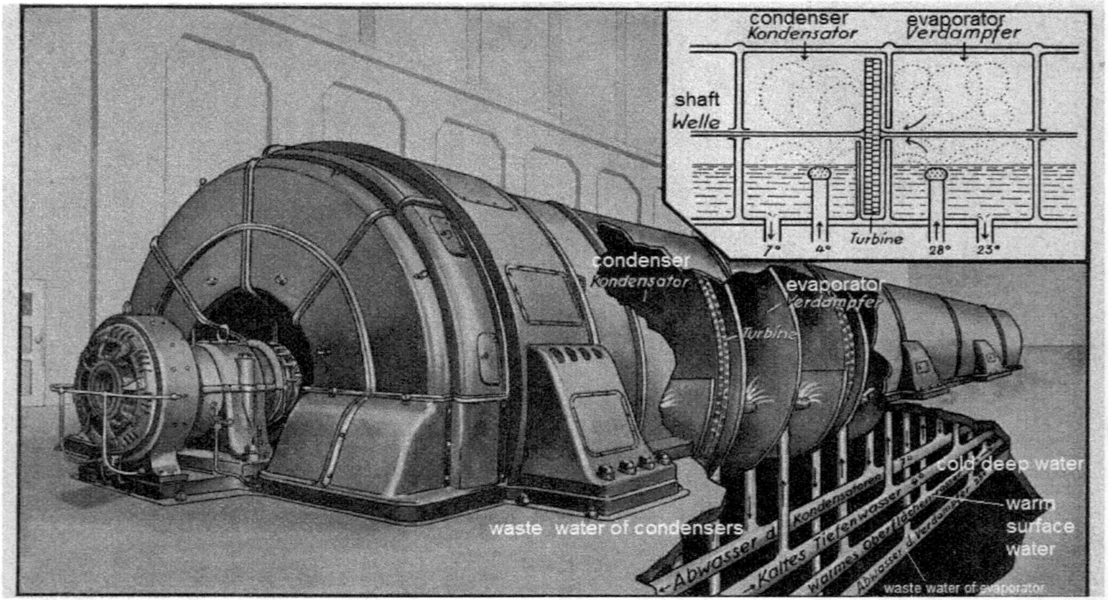

Figure 21. Claude and Boucherot's plan for a turbine generator for sea power plants

Figure 21, caption (continued): The 28°C surface water is sucked in by a centrifugal pump, degassed, and enters the spaces designated as "evaporators", which have been evacuated by an air pump to the evaporation pressure, where some of the water evaporates. The unevaporated water, which is still about 23°C, returns to the sea through a special pipe. The deep water, which is about 4°C, is sucked in by a second centrifugal pump and fed through a special pipe to the spaces designated as "condensers", where it is heated to about 7°C, and then also returns to the sea. Each evaporator and the condenser next to it have a turbine arranged between them, into the center of which the vapor produced in the evaporator is fed; after performing work, the vapor comes back out to the periphery and goes into the condenser, where it is condensed. The turbine wheels have a diameter of 10 to 15 m. The other parts have dimensions corresponding to this diameter.

This is, for the time being, the most plausible assumption, which is also otherwise obvious for Claude, whose work always had a strong chemical orientation. But however that may be, for the general public the only thing that is important is that the desired goal is somehow achieved. The energy available in the form of the warm and cold bodies of water in the oceans is definitely by far the largest energy source on the Earth, after solar radiation and the tidal currents. And this energy source has the special advantage that, in contrast to solar radiation and the tides, it is dependent neither on the time of day nor on the season, and varies only very little even in other respects. Under these circumstances, its development could lay the foundations for a complete reorganization of the industrial and social conditions in the tropics. The reason why is that the plants there constantly produce cold (in the form of the cold waste water) as a "waste product", so not only do they offer inexpensive energy, but also the possibility of cooling work and living space almost free of charge. At a single stroke, this would make the tropics truly inhabitable for Europeans. Thus, from that point on, it would also be possible to use European workers there and industrialize a substantial part of the Earth's surface, where this is now impossible, due to climatic conditions.

Finally, from this perspective, it is also interesting to investigate what places on the Earth are suitable for the construction of ocean power plants. Of course floating plants are conceivable almost everywhere. However, since they involve substantially larger construction and operating costs, stationary power plants are preferable, if possible. In addition to the maximum possible temperature drop in the water, which is only found in the tropics, the coastal sites in question must meet two more requirements: the steepest possible drop from the coast into the cold water layers, and the strongest possible ocean current along the coast, to wash away the quantities of waste water from the power plant. If this is not done, then of course the waste water will gradually cool off the surface water next to it. If this happens, the consequence is obvious: the plant's power will ultimately drop to zero.

If we look at the coastal areas of the tropics where both these conditions are met and where there is simultaneously a corresponding energy demand, then we find suitable conditions above all on the island of Java, Ceylon, Cuba, and the Lesser Antilles, then near Key West and at Miami, Florida, and finally on the coast of Mexico. Even today, all these areas have places where the density of their electric power supply grids and electrical power consumption are completely European. The choice becomes very much larger if we consider selling electric power not only in settlement areas, but rather also another possibility: producing transportable electrochemical products in factories affiliated with the power plant. In this case, it is also possible to consider the coral islands of the South Seas, the reefs of the Sunda Islands, the Maldives and Lakshadweep (Laccadives), the Nicobar and Andaman Islands, the Comoros and Seychelles, as well as the islands in the open Atlantic. They all drop off abruptly to great depths, and they are also all washed by strong currents. However, another precondition comes up here: such places must have a harbor that is accessible to large ocean-going vessels. If that is the case, then such plants are superior to their inland competitors not only because their energy is cheaper, but also in one other very important point: the fact that in principle they lie on the largest highway on the Earth – the ocean – so that the price of their products only has to bear the inexpensive cost of sea freight.

Claude had these possibilities in mind when he prophesied that if his ideas were realized there would be a complete revolution in all existing economic relationships, and the future centers of industry would no longer be in Europe, but rather on tropical beaches. The fact that our economy is so energy-constrained provides ample reason to believe that the inventor was correct in making this prediction.

8. Power from the Arctic Cold

Opposites meet even in technology. Claude's plans for the tropics lead directly to the perpetual ice of the poles, where another inventor also wants to harness a natural temperature drop, namely that between the heat stored in the water under the arctic ice and its surroundings, which are significantly colder. The reason why is that an ice cover forms excellent heat insulation that prevents the cooling of the water beneath it, and the thicker this cover is, the better it does this. Consequently, it is possible, even in the extreme North, to find water under the sea's pack ice whose temperature is 2-3°C above the freezing point, although the mean air temperature in these areas is -22°C. This heat drop between the water and the air in the polar region can, in theory, be used to generate energy in the same way as the temperature drop between the surface and deep water at the equator can. The only prerequisite for this is a substance that is in the vapor phase at about 0°C and in the liquid phase at –22°C, so that it is able, at the temperature range under discussion, to undergo a cycle similar to that of the steam that powers our steam engines between 0°C and 100°C. Few substances meet this requirement; the one that is most suitable is butane, a hydrocarbon compound that under normal pressure boils at –10°C, while it is liquid at lower temperatures. Butane is also not water-soluble, which is a very valuable property for this purpose. These properties of butane form the basis of the project of an "ice power plant" developed by the physicist Dr. Barjot, who wants to use it to supply energy to the far North, above all the Canadian mining areas.

In the previous section we found that we need three devices to exploit a temperature drop: an evaporator that supplies the process vapor, a turbine in which the heat energy of this vapor is converted into mechanical energy, and a condenser that reliquefies the turbine's exhaust vapor. Brief reflection shows that in this case the evaporator can be very simple. Since butane does not dissolve in water and already boils at –10°C, it is sufficient to take the 2-3°C sea water whose heat reserve is supposed to be exploited and mix it with the butane in a closed tank. Then the butane immediately begins to evaporate, while the water freezes in reaction to its heat is being withdrawn. Of course the resulting pieces of ice must be removed; they correspond to the ashes in the ash pan of a coal-fired boiler. Another thing that the evaporator also needs to have is a device to remove the air that is released when the water freezes. – The condenser that reliquefies the butane vapor coming from the turbine can be just as simple: to accomplish this, the vapor is routed into a closed space containing pieces of frozen salt water [19] whose temperature is –22°C. This "salt water ice", which is an excellent coolant due to the large amount of heat required to melt it, very quickly cools the butane vapor so strongly that it is reliquefied. The liquid butane flows directly out of the condenser back into the evaporator, while the brine liquefied by heat absorbed from the vapor flows outside to freeze again in the system provided for this purpose.

The setup and operation of Barjot's ice power plant are shown in detail in Figure 22. The right side of the picture shows the sea with its thick covering of pack ice. Under the ice, an intake pipe opens into the water, whose temperature is 2-3°C. The water is conveyed up through this pipe by means of a pump, and is sprayed into a tank containing liquid butane, which immediately causes it to begin to boil and strongly evaporate, while the water sprayed in freezes and collects on the floor of the evaporator in the form of coarse ice crystals. The butane vapor that

---

[19] Author's note: Of course on the face of it, it would also be possible to use correspondingly cooled liquid brine, since all that is decisive is the temperature drop between it and the butane vapor. Frozen brine is used because the air contained in the brine separates out on freezing, so that it is not necessary to provide an air separator.

is evolved drives a low-pressure turbine that is coupled with a generator, and then goes to a condenser filled with pieces of salt water ice at −22°C, where it is reliquefied as a consequence of the great cold, thawing part of the salt water ice in the process. The liquid butane is fed to the evaporator; by contrast the brine produced from the melted ice is carried outside, where it refreezes.

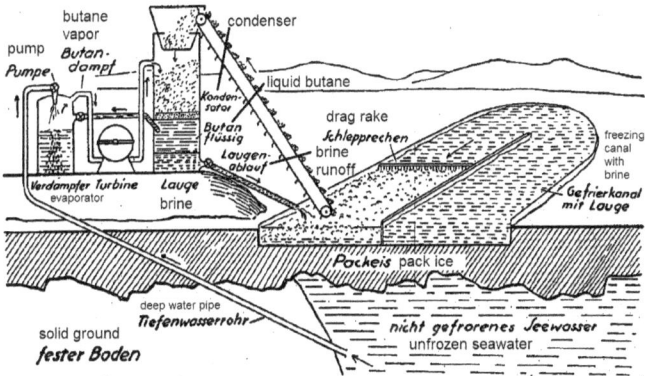

Figure 22. Schematic illustration of Barjot's ice power plant

To accomplish this, Barjot wants to put freezing canals in the pack ice near his power plant, into which the "warm" brine coming from the condenser drains. When it flows out into the cold air it gradually cools off, until it finally begins to freeze under the influence of the low night-time temperature, forming small flakes that get bigger and bigger during the course of the night, but that never combine into a solid layer, since the brine is constantly flowing. These coarse ice crystals are

104

collected from time to time by electric drag rakes
that feed them to a bucket conveyor, which in turn
supplies them to the condenser. The required brine is
obtained by letting sea water freeze until the
remaining salt solution has the right concentration.

Figure 23 supplements the diagram of the entire
plant by showing the details of the boiler and
machine system. On the right, the "salt water ice" is
automatically fed to the condenser by a brine column
cooled to −22°C. To ensure this automatic supply,
the air pressure in the condenser is greatly reduced
by an air pump and constantly kept at this low level.
Consequently, the brine column inside the brine
tower, which is separated into two parts by a
partition, rises considerably higher than outside. On
the other hand, the small pieces of salt water ice
thrown in from outside are lighter than the brine, so
they rise by themselves in the brine tower, without
requiring any assistance. The floor of the condenser
forms a sloping surface, down which the ice slides,
automatically distributing it. The condenser has the
turbines arranged beneath it, and their exhaust vapor
pipes pass through the sloping surface. The butane
vapor flowing out of the exhaust vapor pipes
condenses on the pieces of ice, which form an
extraordinarily large condensation surface, and,
together with the brine produced from the melting
ice, flows down the sloping surface into the
collecting tank on the left end of the condenser,
where the brine collects at the bottom, while the
butane, which has a lower specific gravity, floats on

top. From here, the brine is sucked out by a pump
and fed to the freezing canal, where it freezes again,
while the butane goes into a long container lying just
above the inlet pipe for the water taken from the sea.
The butane container and the water pipe have
numerous nozzles that fit into each other in pairs.
These nozzles feed the liquid butane, in intimate
mixture with the sea water, to the evaporator beneath
the floor of the turbine cluster. The intimate mixing
with the "warm" water causes the butane, which has
been cooled to −22°C in the condenser, to heat up
quickly and evaporate vigorously. The vapor flows
through the short vapor feed pipes leading up into
the turbines, where it performs work. The water, by
contrast, freezes to ice when it is mixed with the
butane, which withdraws heat from the water. Since
these pieces of ice are heavier than the butane, they
sink to the floor of the evaporator, where they slide
down to the left into a collecting tank, from which
they pass under a partition to the outside.

106

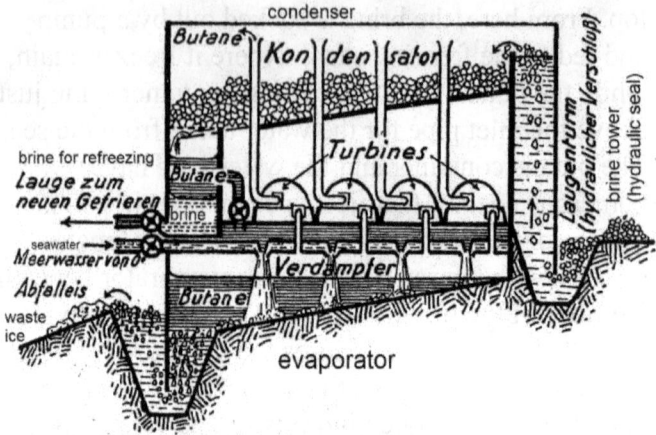

Figure 23. Schematic illustration of the power house
of a Barjot power plant (taken from "La Science et la
Vie")

The power of such a power plant depends first of all
on the quantity of frozen brine that can be produced,
which in turn depends on the size of the freezing
canals. However, aside from this, assuming that the
plant's thermal efficiency is only four percent and
given an air temperature of −22°C, calculation shows
that one cubic meter of water at +2°C can produce
energy equivalent to that of a cubic meter of water
falling from a height of 1,200 meters. To exploit the
water power of such a head would require a very
extensive and costly system. But in the Barjot power
plant the "drop" is in the form of a thermal stress
between the cold and the warm medium, so to speak,
together in the tightest space represented by the
thickness of the separating ice layer. This is an
extraordinary advantage, even compared with

Claude's process, whose two heat sources are separated by a much greater distance, since such large structural means are not required. Accordingly, it is possible to construct a Barjot power plant at relatively low cost. Barjot himself calculates the cost to be only 100 Marks per kilowatt rated power, while hydroelectric power stations cost five to eight times as much. However, the inventor is probably too optimistic about this.

The best places to build such power plants are Northern Canada, Northern Siberia, Alaska, Greenland, Iceland, and the banks of the White Sea, since the longer the region in question has the required low air temperatures, the greater the profitability. Of course this requirement is also met by the Antarctic regions, but there will not be any demand for energy there until the day when the mineral resources that definitely must be present there begin to be exploited. Barjot himself wants to carry out his project in Canada first; the possibilities of doing this are currently being studied. More recent geological explorations have shown that the banks of Hudson Bay and the Arctic Ocean have an area about 5 million square kilometers in size that is extraordinarily rich in mineral resources of all kinds. However, it is currently almost completely undeveloped, since it is not possible to deal with the deep-frozen soil using the means that are conventionally used elsewhere [20]. How quickly this state of affairs would change if inexpensive current were available there, i.e., inexpensive power, inexpensive light, inexpensive heat. The situation is similar for the Mackenzie Basin with its large petroleum reserves, for the gold and the zinc ore near the Great Slave Lake, for the iron ore on the Belcher Islands, and the enormous deposits of high-quality copper ore on the Coppermine River east of Great Bear Lake at 67° north latitude. Northern Canada, which is virtually covered with lakes, also offers favorable locations for Barjot plants in the country's interior. Not only that, but especially during the few summer months when the ice power plants in the interior would, for obvious reasons, suffer from a decrease in energy, the melting of the ice in the rivers and waterfalls would compensate for this by making available incalculably large amounts of water power. Coupling these two possibilities would ensure that even the greatest energy demand could be covered throughout the entire year without interruption, and thus allow the development even of regions that have been considered completely uneconomical up to now.

However, in other respects the operation of Barjot plants is in no way limited to the polar regions, since even lower latitudes have temperatures down to −18°C for longer periods of time. E.g., in the North

---

[20] Author's note: E.g., in the Klondike gold hunters can only mine the gold deposits lying close to the surface if they first thaw the ground with jets of steam.

American industrial areas between Hudson Bay and the Great Lakes, such temperatures are not a rarity during most of the winter; here coupling with hydroelectric power plants during the summer months would also be possible. Finally, however, it is also possible to operate the plant at a somewhat higher outside temperature if the frozen coolant is stored or using a lower-concentration brine that freezes at a higher temperature.

In any case, the overall result is that the Barjot process is of the utmost importance for supplying the very coldest regions of the Earth with energy, a task which has hardly been addressed so far. Moreover, it might later assume the same role that steam power reserves now play in the winter for hydroelectric power plants at lower latitudes: making up for the immensely increased demand for electricity in the cold season, when hydroelectric plants have less power because of the decreased quantity of water.

Figure 24 shows roughly what such a plant would look like to a viewer. Here the place of the bucket conveyor is taken by a conveyor belt that puts the pieces of ice into a reservoir. We might possibly already see such plants set up even in the near future, since the mineral resources of the Arctic have long been crying out for development, which requires energy.

110

Figure 24. The polar world should also supply energy in the future: This shows a power plant project that uses the Arctic cold to generate energy.

9. What about Geothermal Power?

This question is obvious after hearing that the heat drop of the seas can be used for energy generation, since everyone knows that there is also an abundance of heat in the Earth. Volcanoes are living proof of this. This geothermal power is a dowry that the Earth received when it separated from its mother the Sun millions of years ago. The Earth was then a fiery liquid, probably first a massive ring that then collapsed and formed into a ball. As this ball slowly cooled over the course of time, it became covered on the outside with a crust that became more and more solid. Scientists assume the thickness of this crust to be about 50 kilometers. We know nothing from experience about what lies below it. However, it is suspected that at a depth of 200-300 kilometers the inside of the Earth is still a red hot liquid even today. This opinion is based above all on the fact, repeatedly experienced when holes are drilled and tunnels constructed, that the temperature of the ground very quickly increases with the depth by an average of 3°C for every 100 meters (the geothermal gradient). This gradient means that at a depth of 4,000 meters the temperature is already about 120°C, a temperature at which all water that is present boils, despite the higher air pressure. Thus, the old joke about the large furnace that humanity is sitting on is perfectly correct in principle.

This brings us back to the initial question of whether it might be possible to exploit the heat of this "furnace". Of course, it would be most favorable if it were possible to get directly to the innermost magma chamber, since at 5,000°C its heat energy has been estimated at a nonillion kilogram meters. This number is so enormous that no one can even conceive of it. However, from the technical perspective it is of no interest, since we have no way of mastering this heat and also no way to approach it. However, this does not mean that geothermal power is completely inaccessible, since the Earth's c r u s t also contains natural "furnaces", those small pockets of igneous magma whose vents form volcanoes. Their heat would be easier to use.

Just two decades ago, this idea would have seemed a total fantasy, absurd, impossible, a utopia that not even Jules Verne imagined. Today the problem is already being seriously discussed, since in Italy during World War I a power plant was built whose entire output of over 12,000 kilowatts came from volcanic heat in the form of hot steam that was used to heat its boiler. This first volcanic power plant is in northern Tuscany, at the village of Larderello near Volterra. The entire region is mountainous and bears a strong resemblance to the Eifel mountain range in western Germany. However, to the visitor who is unacquainted with the region, Larderello itself appears to be a little piece of Hell that has become visible, since the mountain slopes there, which are devoid of any plant growth, show hundreds of cracks and fissures everywhere called *soffioni* that have highly heated steam roaring and hissing out of them. For over 100 years this steam and also the hot springs (*lagoni*) coming to the surface in the same area have been exploited for their boron content by the Società Boracifera di Larderello to prepare boric acid. In the process, in 1904 someone hit upon the idea of using the thermal energy of the of steam to drive power machines. The first attempt involved a 40-horsepower piston steam engine. It helped operate not only the machinery of the boric acid factories, but rather also a small lead rolling mill and later an electric utility that provided the factories and the village with electric light. The entire plant worked perfectly satisfactorily until one fine day the steam engine simply stopped working, since its internal parts were completely corroded. In the joyful exuberance about the kindness of Nature, which was apparently bestowing handfuls of presents, one minor detail was overlooked: that the steam's high content of borates, ammonia, traces of sulfuric acid, etc., is an additional gift that is not a very good for the operation of steam engines. This experience was bitter, but its lessons were learned. The company had already long intended to use the steam of the *soffioni* on a larger scale to produce electrical energy, for which ample demand could be found in the surrounding cities. To realize this plan, it was first necessary to rebuild the entire system on a sound foundation. Careful consideration showed that the goal that was sought could only be achieved indirectly: by using the natural steam to heat an evaporator that produced pure process steam from pure water. This was also the approach that was pursued, and this was first done with an experimental plant opened in 1912 that contained a specially built standing water-tube boiler and a 300-horse power steam turbine with directly coupled generator. The results were favorable. Therefore, in view of the strong demand for electric power in the military industry, right at the beginning of World War I the company decided to use it as a basis for building a larger, long-distance power station as quickly as possible; it was opened in 1916 with three generating sets, each having power of 2,500 kilowatts, and in the meantime it has been expanded to have a total power of 12,000 kilowatts. The current that is produced goes through five long-distance lines into the networks of the cities of Volterra, Siena, Livorno, Cecina, and Florence, where it supplies the streetcar system, among other things.

Since tapping the natural steam sources usually presents difficulties, the plant's heating steam is taken from 40 centimeter-wide boreholes sunk specifically for this purpose. In general, the boreholes are 60-120 meters deep; however in isolated cases they have even gone down to 150 meters. During the work, the borehole is lined with welded iron pipes. Once it has reached the right depth, the drilling probe is removed and replaced by a roughly fitting piston, which is driven down as far down as possible, and then pulled out very quickly with an electric winch. This produces a partial vacuum in the borehole, allowing the steam in the ground to break through the thin layer of soil that is still present. The first consequence is small, volcano-like eruption of mud, rocks, steam, and water, which is often so powerful that it smashes the drilling rig. A few minutes later everything is over, and now there is a peaceful, steady outflow of dry steam having a mean temperature of 180°C and a mean pressure of three atmospheres above atmospheric pressure. Many years of experience have found that this pressure remains completely constant; there is just as little change in the productivity of the steam sources, which is 3,000-14,000 kilograms per hour, depending on the location.

Once the borehole has been completely cleaned out, then it is connected with a collecting line that carries the steam of all bore holes to the boiler house. The apparatus there is shown in Figure 25. We see that the natural steam first flows through a superheater and then an evaporator, where part of it condenses to water. The rest of it, along with the gases it contains, goes through the exhaust steam line into the production plants, where carbonic acid and recently also helium are extracted from it. The almost pure water, whose temperature is about 90°C, condensed from the natural steam is also used: It feeds the evaporator, which works like a water-tube boiler and consists of about 300 aluminum tubes seven meters long in a sheet metal jacket. The tubes filled with condensed water have hot natural steam flowing around them. This causes the water to boil and converts it into steam, which first passes through a water separator to dry it and then through the superheater, which is also heated with natural steam. From here it is fed in the form of hot steam to the turbine, which turns the generator that is coupled with it. The exhaust steam of the turbine is condensed to water in a surface condenser, and some of it is returned to the evaporator.

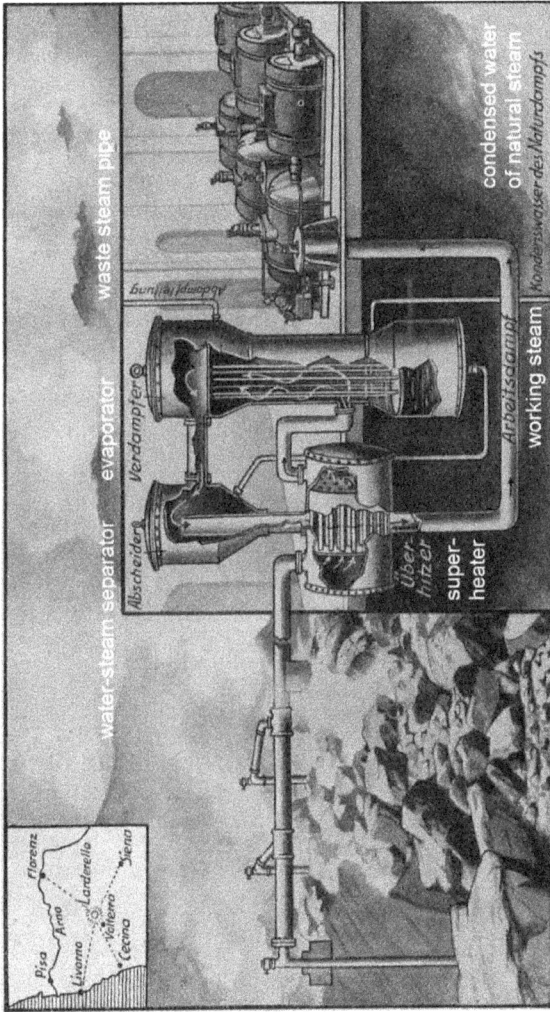

Figure 25. The components of the Larderello volcanic power plant, which today has a power of 12,000 kilowatts and supplies five cities with electric current.

As was mentioned, today the power of this volcanic power plant is already 12,000 kilowatts. This is not trivial, but it is only just a beginning, and it has been proposed to expand it soon, if experience continues to be favorable. This should be kept in mind when considering the news that it has recently been decided to drill substantially deeper holes in the hope of developing significantly greater steam sources. The first success has already even been achieved: On March 26, 1931 a *soffione* was drilled whose power far exceeded the maximum known up to then. At a depth of about 360 meters the drilling probe broke through the cap of the underground dome, whereupon a 300 meter high steam jet loaded with water, mud, and rocks shot out of the borehole with a deafening roar could still clearly be heard at a distance of 50 kilometers. According to preliminary calculations, this borehole has an hourly output of 260,000 kilograms of 160°C steam, while the total hourly steam consumption of the current power plant at full load is 168,000 kilograms. Thus, the new steam jet by itself can supply a plant having a rated power of 15,000 kilowatts.

This small calculation shows better than many words how much energy can be generated by methodical development of the underground boiler heated by the fire of volcanoes. The reason why is that there are many other areas on the Earth that are also blessed with natural steam. Some of these steam sources would be even better for energy generation than the *soffioni* of the Tuscan Maremma, since their steam is free of harmful constituents.

This is the case with the fumaroles in the area of Lago, which also belong to the Società Boracifera di

Larderello. A 10,000 kilowatt power plant is currently being built there that is supposed to use the natural steam directly in turbines. Of course such direct utilization greatly increases the overall efficiency. The process was tested for several years in a small pilot plant with good results.

Furthermore, extensive preliminary work has been undertaken in the area of the solfataras of Pozzuoli (in the province of Naples) to clarify the practical possibilities for exploitation. This would involve direct utilization of the heat of Mount Vesuvius. If this is accomplished, then Mount Etna would be the next point of attack. These efforts have special importance for Italy, since the country does not have any anthracite or bituminous coal, and only small deposits of brown coal, while the relatively rich water power is limited mainly to Northern Italy. Supplementing it with Southern Italian volcanic power plants would save Italy a large part of its coal imports.

But things are also happening in other places, above all in the United States and on the island of Java. About 60 kilometers north of San Francisco lies a mountain range that is extraordinarily rich in extinct volcanoes, the Mayacamas. Extending along its west flank is a large tectonic fault line along which natural steam sources come to the surface in numerous places. Obviously, close to the surface there is an old magma chamber whose hot gasses heat the entire area so that the water often boils at a depth of just half a meter. The first attempt here to drill steam sources producing a regular supply of high-pressure and high-temperature steam was undertaken by the General Electric company in the summer of 1921. Since then, several boreholes have been made, some of which supply very large quantities of steam with temperatures of 150-190°C under pressures of up to 13 atmospheres. Industrial use has been disregarded for the time being. Only one small turbine set of 35 kilowatts was set up for the purposes of lighting and to operate the drilling installation. Aside from that, it was considered sufficient to make regular measurements to find out whether the quantity of steam really remains constant when allowed to escape freely. So far, this expectation has been met: the output does not show any sign of fluctuations worth mentioning. Even setting up a borehole in the immediate vicinity of an already existing one does not have any effect on steam production. Under these circumstances, exploitation should begin soon.[21] In order to get a precise idea about the possibilities for doing this, a committee has already been sent to Larderello to study this question.

On the island of Java, similar experiments were undertaken in 1926 in the fumarole area of Kawah Kamojang in the Guntur volcano group. The Dutch geologists had a very favorable opinion about the

---

[21] Translator's note: A geothermal power plant has been in operation there since 1983:
https://www.energy.ca.gov/powerplant/steam-turbine/sonoma-geothermal-project-unit-3

future of the undertaking, since the mean steam production here is so large that each borehole can be expected to produce about 900 kilowatts. Based on this figure, even if the invested capital is depreciated as quickly as possible a large volcanic power plant set up here would produce current at a price of only 0.2 to 0.3 pfennigs per kilowatt hour, while the production costs per kilowatt hour at a large hydroelectric power plant nearby are more than four times as much.

It can be seen that a good start has been made in this area. And there is no doubt that others will also do the same in other places, as soon as further successes eliminate the remaining reservations. Japan, New Zealand, fuel-starved Iceland, Alaska with the Valley of Ten Thousand Smokes, probably the largest fumarole field on the Earth – all of them offer especially favorable preconditions for the development of the new energy source under discussion here. But one should not think that only volcanically active areas allow exploitation of geothermal power. They are the outposts that have to collect the first experience. However, the ultimate goal must be to put geothermal power at our service quite generally and everywhere. This idea also has already been expressed, and there is even a project for it whose creator is no less than Sir Charles Parsons, the inventor of the steam turbine. He proceeds from the phenomenon, which is well known to geologists, that even in countries where volcanic activity is completely extinct there are places where the Earth's heat increases not by only 3°C per 100 meters depth in the ground, as it does in other places, but rather much more quickly, so that when drilling quite high temperatures are already encountered at a relatively small depth. At such places Parsons wants to use all available technical means to drive down shafts far enough to reach a temperature of about 200°C. This would normally be at a depth of 6-7 kilometers. At each site, Parsons needs two such shafts that he proposes to line with steel pipes and connect down in the Earth via large hollow spaces. Everything else would then be simple. The first shaft constantly supplies water from above, which is quickly heated on its way down and at a great depth finally turns to steam; the steam escapes through the second shaft, where it is fed to large turbines. Basically, Parsons intends to use engineering to make places where volcanic activity has long been extinct produce what Nature gives away free in volcanic areas. He intends to produce real boilers in the Earth, all of them heated by harnessing the fire of the Earth's core. The energy that can be generated from such a system ultimately depends on the quantity of water drained in every hour, whose drop would also, incidentally, be expedient to exploit in the upper depth levels. Hydroelectric power stations that draw ten cubic meters of water per second are no longer a rarity today. Warming this quantity of water to 160°C produces power of around three million kilowatts; a normal steam power station would have to burn 70,000 tons of coal to generate this much power. We calculated the world's energy demand in 1970 to be one billion horsepower (see page 11). That is around 750 million kilowatts. Thus, 250 geothermal power plants of the indicated power would be more than enough to cover the world's entire energy demand for many decades. This is the best illustration of what enormous energy supplies are hidden here.

The only problem is that for the time being we are unable to tap them! This is because our engineering capabilities are nowhere close to adequate for boreholes 6-7 kilometers deep. And at such depths we are even less capable of excavating hollow spaces, which are necessary to connect the two boreholes. These and other objections, for instance the low thermal conductivity of the rock, have been raised against the plan to show it to be impossible. But here again, these objections overlook the fact that this task only has to be solved in the still distant future, when the means of engineering will certainly be quite different from those that we have today. A hundred years ago, who would have considered it possible that divers would one day search for sunken treasures at a depth of 200 meters in the sea? In another hundred years why should it not be possible to drive shafts to depths that are inconceivable today and, at the bottom of them, have artificially cooled automatic drilling and excavation machines working at temperatures of perhaps 160°C?

However, there is also another variation of this project without the "boilers" deep down in the Earth, using artificial furnaces instead of natural ones: an idea that Sir William Ramsay, known as the discoverer of the noble gases, published in 1914 following the conclusions of the geological congress in 1913, which served as the starting point of our discussion. Ramsay said to himself that the simplest way to avert the impending coal shortage would certainly be to tap the many coal deposits that we cannot exploit today, since they are too thin to be extracted by mining. This is the case for all seams less than 80 centimeters thick, and the practical result of these conditions is that almost three quarters of all known coal reserves on the Earth are found in such seams (see p. 5). Ramsay's suggestion to tap these quantities of coal that are inaccessible to us today was modeled on salt mining. There are also very thin deposits of salt in many places that cannot be extracted by mining. In these cases, water is pressed down through boreholes to dissolve the salt and then pumped back up, and the salt is recovered from the brine by evaporation.

We cannot dissolve the coal, since there is no known solvent for it. But it could be ignited in places using electricity, and the burning coal deposit could be supplied with compressed air and steam in alternation through a pipe. The compressed air would blow the fire to white heat. In the process, the coal would burn to carbon monoxide, which the subsequent steam converts to water gas or mixed gas. This mixed gas, which is superbly suitable for operating gas-powered machines, and just as good for heating purposes, would be sucked out of the burning seam through a wide pipe, cooled, purified, and collected in large storage tanks. Figure 26 shows roughly how such a coal gasification plant would look.

Figure 26. Ramsay's plan for gasifying unmineable
coal seams inside the Earth

The gas produced could be used to run gas-powered
machines and generators. Our large smelting plants
have had such gas-powered machines for a long

time, with the only difference that they are powered not by mixed gas but rather by blast furnace gas. However, another combination would be preferable, to which we are led by a question that is very obvious in this connection. Does the generation of electricity always involve a power machine and a generator? Are there no other ways to achieve this goal? For example, is it possible to convert heat or light directly into electricity? The answer is that such conversions are already possible, but for the time being only in the laboratory.[22] That is, all the engineering here still remains to be done. What basis there is for this will be shown in the next chapter.

10. The Ultimate Goal

The question of direct conversion of heat into electricity has already occupied science and technology for over 100 years – since 1821 when S e e b e c k discovered the phenomenon that physicists call "thermoelectricity" for short. Many readers might still remember doing the basic experiment in school. Two strips of different metals are soldered together on one side, and the point

---

[22] Translator's note: The French physicist Antoine-César Becquerel discovered the photovoltaic effect in 1839, but it was not until 1941 that Russell Ohl developed the silicon solar cell, and solar cells with efficiencies of over 20% were not developed until the 1980s.

where they are soldered is heated. The result is an electric current that can be tapped at the unsoldered ends. So far, everything is fine, however, there is a "but". To this day, no one knows what is going on inside the metals when this type of current is produced. Consequently, any possibility of directing the process according to our wishes escapes us. The tests that have been done on such arrangements so far show an efficiency of perhaps one percent in the most favorable case. Thus, only $1/100$ of the input heat energy is utilizable. This precludes any large-scale use of it. Fortunately, however, the thermocouple is not the only way to achieve this goal. Science has already known for a long time that the coal oxidation process customarily referred to as "combustion" can also be conducted in such a way that the energy released by it appears not as heat, but rather as electricity. This "cold combustion" can, under favorable conditions, have an efficiency as high as 95 percent. The process is referred to as "cold combustion" because the reaction of coal with atmospheric oxygen takes place entirely without formation of heat or even flame. By contrast, the expression does not imply that the process takes place at a normal temperature, since the reaction between coal and oxygen is only possible at relatively high temperatures. "Combustion", the consumption of coal, is the fundamental difference between the f u e l   c e l l based on this principle and the known galvanic elements, whose most important representative is the dry cell of our telephone wires. In the dry cell, carbon also forms one electrode, while the other consists of zinc. But what undergoes "combustion" in the dry cell is the expensive zinc, while the carbon remains unchanged.

The first person to tackle the problem of building an element that burns inexpensive carbon, instead of an expensive metal, was the French chemist Becquerel, the discoverer of radioactivity. It has already been 80 years since he conducted his experiment. His tracks were later followed by other researchers, also including the Swiss Emil Baur, who after lengthy investigations obtained a very high yield with relatively simple means, so that in practice his process is by far the most promising. Baur's fuel cell consists of a large fireproof box that is kept at a temperature of about 1,000°C by outside heating. This box contains molten copper, which has a thick layer of copper oxide (copper slag) floating on it that is also molten. The liquid copper has an iron tube submerged into it, through which air is blown. This tube simultaneously forms the positive pole of the cell, which supplies direct current, while the negative pole is formed by a block of carbon submerged into the copper slag, but not touching the copper layer. As soon as air is blown into this arrangement, the following process takes place: The atmospheric oxygen combines with the copper to form copper oxide, which has a lower specific gravity, so it rises to the top, where it touches the carbon. The high outside temperature then causes the carbon and copper oxide to react with one another in such a way that the carbon is burned to form carbon monoxide, while the copper oxide is reduced back to copper. This reaction releases electrons, which exit in the form of electrical current.

The entire process consumes only carbon and oxygen; the copper converted into copper slag is always regenerated. However, the reactions have not yet gone to completion. The reason why is that the combustion of the carbon (C) to carbon monoxide (CO) does not represent complete utilization of the energy contained in the carbon, since each atom of carbon can bond with not just one atom of oxygen, but rather two. Therefore, Baur immediately passes the carbon monoxide that is produced back into a second system having a somewhat more complicated design, where it is burned to carbon dioxide (CO_2) at an outside temperature of about 800°C. This process is also carried out in such a way that no heat, but rather only electricity is produced. Of all the energy in the carbon that is consumed, no less than 60 percent is made available. Compared with the energy balance of the circuitous route conventionally used today: "boiler – steam turbine – generator", this represents quite extraordinary progress. [23]

This opens prospects with which we could be perfectly satisfied, if the whole process did not involve an extremely costly factor: the necessary of constantly keeping the container a temperature of

---

[23] Author's note: In the most favorable case, the economic efficiency of the most modern steam engines is 30%, while a modern generator an economic efficiency of has 96%.

about 1,000°C. This precondition is the main reason why 40 percent of the promised 100 percent theoretical efficiency is lost. This changes if we combine the idea of the fuel cell with Ramsay's bold suggestion of gasifying the coal seams in the Earth that cannot be mined: Where such facilities exist, the fuel cells are heated with mixed gas. This kills two birds with one stone: the immense coal deposits that cannot be mined are exploited, and cheap heat is created to operate the fuel cell power stations. Such electric power stations would no longer have any smokestacks, no smoke, no soot, no boiler houses, no turbines, and no generators. Wide halls would have heat-resistant containers standing in them as tall as a house; these containers would hold molten copper and be heated with gas, similar to a Siemens-Martin furnace, and be connected through enormous tubes with electrically operated compressors that constantly blow air into the melt. The coal that is consumed would be replenished by constantly blowing in coal dust. This would make the entire operation practically completely automatic. A few engineers would be sufficient to monitor it.

However, the question still arises whether in a future that has the technical ability to master such tasks we will not also be so much further along than today that even "cold combustion" is a detour. Probably by then all the physical prerequisites of the t h e r m o couple will long have been known, so that it will be possible in this way to convert the heat of the Earth directly into electricity. However, perhaps by then we will also already be capable of producing electricity directly from solar radiation using photoelectric effects, since we can already do this on a very small scale today. The most recent progress in this area is connected with the name of the German physicist Dr. B. Lange. To accomplish this, he built a small apparatus consisting mainly of copper plates and copper oxide layers, which can work under normal air pressure, in contrast to older photocells. If the set of plates is irradiated by the Sun, then chemical reactions take place in it, which are accompanied by electric currents. The efficiency of this procedure is also not limited the way it is with a heat engine. Instead, the principles of thermodynamics would allow a theoretical efficiency of no less than 95 percent, as in a fuel cell. But here also we must wait for the future, since for the time being these are only laboratory experiments.

From today's perspective it is even more difficult to overlook a third problem of physics, the greatest that engineering physics still has to solve: utilization of the energy inside the atom by artificially splitting atoms. Einstein has shown us how to calculate the energy value of matter. According to his result, a single drop of water contains enough energy to supply 200 horsepower for an entire year. Accordingly, the values involved are, relative to the same weight, a hundred thousand, or even a million times greater than those we get from combustion and other chemical reactions. The only problem is that for the time being we do not have the ability to reach them or to unleash them. In radioactive substances we can watch how the energy inside the atom is released in the natural way. But these substances are either too rare or too inert to allow their technical exploitation for the purpose of power. And in the places where we are able to bring about such atomic conversions ourselves – for example the splitting of the aluminum nucleus – the yield of split atoms is much too small to motivate the engineer. Some years ago it seemed as if physics at least had a handle on the great secret, since at that time the conversion of hydrogen into helium was reported. But the news turned out to be false. Thus, this area also still remains closed to the engineer for the time being, until science one day gives him the key. Today no one can say whether years or generations will pass in the meantime. However, one day humanity will certainly also reach this goal. Then it will smile as it looks on the powerful hydroelectric and thermal power plants that today supply the Earth – still so inadequately – with electricity. And instead of satisfying the hunger of enormous boilers with delicacies like coal and oil, or building dams for countless millions, it will prescribe a diet of atomic energy to novel miniature machines, with the result that their power increases to many thousands of times that of which their ancestors were capable. When this day comes, everything that today serves for the transport of fuel from one end of the Earth to the other will disappear, since the annual fuel requirement of an electric power plant will fit in a water bottle, and all work will quite obviously be done everywhere with electricity. If we think through these possibilities to their ultimate conclusion, then we see before our eyes a colossal energy vision that by far exceeds the boldest wishes of today's technology. Humanity will then have unlimited amounts of energy available, which will be just as suitable to sow death and destruction as life and happiness. If humanity is not mature enough for this power when the future lays it in its hands, then peoples and countries will be destroyed by it. They will be victims of "persons equal to gods", who really swing lightning in their fists.

Index

* 9 7 8 0 9 9 8 7 8 8 7 8 4 *

www.ingramcontent.com/pod-product-compliance
Lightning Source LLC
Chambersburg PA
CBHW071706210326
41597CB00017B/2360